Tensegrity Systems

Robert E. Skelton • Mauricio C. de Oliveira

Tensegrity Systems

Springer

Robert E. Skelton
Mechanical & Aerospace Engineering
University of California, San Diego
9500 Gilman Drive
La Jolla, CA 92093-0411
USA
bobskelton@ucsd.edu

Mauricio C. de Oliveira
Mechanical & Aerospace Engineering
University of California, San Diego
9500 Gilman Drive
La Jolla, CA 92093-0411
USA
mauricio@ucsd.edu

ISBN 978-0-387-74241-0 e-ISBN 978-0-387-74242-7
DOI 10.1007/978-0-387-74242-7
Springer Dordrecht Heidelberg London New York

Library of Congress Control Number: 2009921739

Printed on acid-free paper

Springer is part of Springer Science+Business Media (www.springer.com)

To Judy, who inspires me,
and to my many students who dared to do something different

Bob

To Dani, Bea, and Victor

Maurício

Contents

Preface

The purpose of this book is to make contributions to the analytical machinery required to integrate structure and control design and to show that this optimized structure usually has a finite, rather than an infinite, complexity. The first challenge in such an endeavor is to choose the right paradigm for structure design. Engineers currently design wings for aircraft that coordinate the aerodynamics and structural design disciplines very well, but then the control functions are added, almost as an afterthought, destroying the flying efficiencies that were so carefully treated in the structure design in the first place. The added control functions are not efficiently performed either, given that the structure design has already been completed and cannot be economically redesigned to "cooperate" better with control functions. So, historically, engineers design structures first (beams, plates, and shells) and then add a control system to "torture" the structure into doing things it was not originally designed to do. Even in research communities we have not solved this problem of integrating structure and control design. As a recent research example of this fact, DARPA sponsored a "smart structures" research program to control wings by warping them (as the Wright Brothers did many years ago). An existing F16 wing was instrumented with a nickel–titanium actuator to apply a large torque to the wing to twist it, trying to obtain a 20° twist. The control system and the actuator twisted the wing only 7° because too much power was required to twist the structure away from its equilibrium. As opposed to a control system that torques or pushes the structure away from its equilibrium, we will show that a tensegrity paradigm for structures will allow one to *modify the equilibrium* of the structure to achieve the new desired shape, so that power is not required to hold the new shape. Integrating structure and control design will require less power from the control system to accomplish the same objectives.

Less control power also impacts the parameters of structure design, since less structural stress is imparted to the structural components during control. Such cooperation between the static and dynamic properties of the structure and the control system can only be accomplished by a structure design paradigm that maintains an extremely high degree of "controllability" during all phases of the structure design. This requires new types of structures and the tensegrity structural paradigm is the only one the authors have found with these properties.

This book focuses on the development of analytical tools that would allow the dynamics and structure design to have the control problem in mind, but analytical tools first have to be derived for the static and dynamic analysis of the tensegrity paradigm, leaving room for only one chapter on control.

Even though this book focuses on *mechanical properties* of a tensegrity system, one need not limit design goals to merely mechanical properties. Other books will be written to optimize the geometric arrangement of material (within the tensegrity paradigm) to obtain a prescribed set of other *acoustic properties, or electromagnetic properties.* There are material properties that can be obtained by a "composite" design (where one chooses materials and the topology of their configuration) that natural materials do not possess. For example, we have demonstrated in the microwave lab a "left-handed" tensegrity material, meaning a material whose electromagnetic properties have negative values of permeability μ and permittivity ϵ.

Chapter 1, Introduction and Motivation, will give many small examples to illustrate the basic principles and the fundamental rules that govern tensegrity systems. Examples show tensegrity in nature (at both the nanoscale and the mesoscale), tensegrity in art, tensegrity in architecture, and tensegrity in engineering and science. This chapter is a guide to the book, referring to the types of problems to be addressed in more detail in later chapters and giving some of the optimization results that are simple to explain without the mathematical derivations that follow in later chapters. The rest of the book is fairly dense with mathematics, but Chapter 1 avoids math, to enlighten the spirit and intent and goals of the new methods.

Chapter 2, Analysis of Static Tensegrity Structures, provides the analytical tools to deal with the question *How many ways can sticks and strings be connected to yield a stable equilibrium*? This chapter finds the entire set of admissible member forces associated with any given configuration of the tensegrity system. The chapter also computes the stiffness matrix, predicts the stable configurations, and develops tools for kinematics.

Chapter 3, Design of Compressive Structures, is the first chapter on design of tensegrity systems. The chapter provides algorithms to minimize mass under compressive loads, subject to yield and/or buckling constraints, yielding minimal mass tensegrity structures. The molecular structure of the spider fiber, which is nature's strongest tensile material, is indeed a class 1 tensegrity structure. Algorithms are given to design self-similar structures in an optimal way, to minimize mass, subject to buckling constraints. Formulas are also given to optimize the complexity of the structure, which turns out to be finite. After each additional self-similar iteration, the number of bars and strings increases, but, for a certain choice of geometry, the total mass of bars and strings decreases. This algorithm generates what we call *tensegrity fractals.* For certain structures, the string mass monotonically increases with iteration, while the bar mass monotonically reduces, leading to minimal total mass in a finite number of iterations. The number of iterations required to achieve minimal mass is given explicitly in closed form by a formula relating the chosen geometry and the material properties.

Chapter 4, Design of Bending Structures, repeats the steps of Chapter 3, except for structures under bending loads. This chapter shows that the minimal mass structure, subject to bending loads, is a class 2 tensegrity structure. This work also shows how to optimize the complexity of a structure. The 1904 work of Michell derived the infinitely complex case, a continuum. Indeed, if mass at the construction joints is ignored, then the design of our optimal structure of finite complexity approaches the design of Michell, in the limit as the complexity of our design approaches infinity. If a mass penalty is associated with the presence of joints then the optimal complexity is finite.

Chapter 5, Analysis of Tensegrity Dynamics, derives the dynamics of tensegrity structures. Tensegrity concepts have been around for a long time, but in the absence of efficient algorithms to compute the dynamics, engineering designs were a rare novelty. This might explain the long gap between the invention of tensegrity as an art form and tensegrity as an engineering method. The large number of components, bars, and strings left many convinced that the required engineering would just be too complicated. But we must remind ourselves that "complication" is often measured by the relative ease with which the user of the analytical tools can do the design. Efficient new algorithms are now available to automatically, with minimal labor from the user, generate the equations of motion. This can be an enabler to engineering progress with tensegrity concepts in engineering and science.

NASA spent a lot of money in the 1960s and 1970s developing efficient algorithms for computer simulations of *multibody* systems. Each body in the system could be rigid or elastic, and many applications were forthcoming in spacecraft control, such as in SKYLAB. While these algorithms for generating dynamic equations of motion could indeed be used today for tensegrity systems, Chapter 5 will develop an even more efficient method for describing the dynamics. System constraints are added in the case of structures pinned to ground, or having fixed equal-length bars. This new result presents the dynamic equations in the form of a matrix differential equation in lieu of the traditional vector differential equations. Even though these equations are written using a non-minimal set of coordinates, efficiency comes from the very simple mathematical structure of the equations.

Chapter 6 will focus on methods for controlling tensegrity systems. Some recent results on the control of nonlinear non-minimal systems are derived to allow large motions of tensegrity systems. These methods are based upon Lyapunov stability theory. There are some new control-theoretic results here, but the main focus is to demonstrate the controllable features of tensegrity structures.

This text was designed for two different types of readers in mind. A person who wants the results without trudging through the proofs can read the chapters and skip the *Advanced Material* session at the end of each chapter. Those readers seeking a more rigorous treatment can read the chapter and may appreciate the more advanced material.

Finally, the authors would like to thank and acknowledge the many students and visitors that have been involved with their research on tensegrity in the past. This book mentions the work of Jack Aldrich, Joe Cessna, Waileung Chan, Carlos Cox, Bram de Jager, JeongHeon Han, Milenko Masic, Tino Mingori, Jean-Paul Pinaud, Cornel Sultan, Jeff Scruggs, Carlos Vera, Darrell Williamson, and Anders Wroldsen, among others.

Throughout the text, many plots and figures are better understood and appreciated in color. We tried to generate these figures in such a way that the most important information could be captured from the black-and-white prints of this book. For completeness however, in many points of this book, the reader might encounter references to colors in the figures. In this case, please refer to the full color version of these figure which are available in the website (www.springer.com) or on the electronic version of this book.

Chapter 1

Introduction and Motivation

Buckminister Fuller [Ful59, Sad96] coined the word *tensegrity* as a conjunction of the two words *tension* and *integrity* [Lal96]. Our interest is in engineering structures. Hence, to make progress for any precise mathematical work, the definition of the word tensegrity must be sharpened.

The structures of interest have compressive parts and tensile parts. At the moment, we will label the compressive parts simply *rigid bodies*, and we will label the tensile parts simply *strings*. The set of positions and orientations of all rigid bodies will be called the *configuration* of a set of rigid bodies. Strings may be connected to each other and to the rigid bodies. We refer to set of connections between rigid bodies and strings as the *connectivity*. We shall distinguish the set of connections performed via strings, i.e. the *string connectivity*. Our definition of a *tensegrity configuration* of rigid bodies follows.

> In the absence of external forces, let a set of rigid bodies in a specific *configuration* have torqueless connections (e.g. via frictionless ball-joints). Then this *configuration* forms a *tensegrity configuration* if the given configuration can be stabilized by some set of internal tensile members, i.e. connected between the rigid bodies. The configuration is not a tensegrity configuration if no tensile members are required and/or no set of tensile members exist to stabilize the configuration.

In other words, we say that the configuration of rigid bodies is a *tensegrity configuration* if there exists a *string connectivity* able to stabilize the configuration.

We distinguish between the internal forces acting on the rigid bodies from string attachments and the *external forces*, i.e., forces that do not come from strings within the structure. Of course, there exists a set of external forces

DOI 10.1007/978-0-387-74242-7_1, © Springer Science+Business Media, LLC 2009

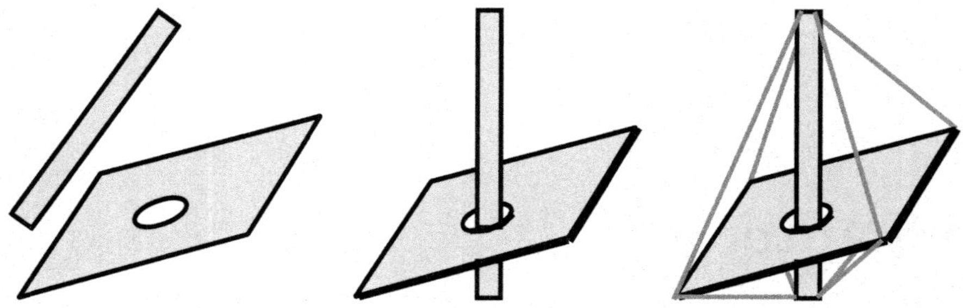

(a) Not a Tensegrity Configuration (b) Tensegrity Configuration (c) Tensegrity System

Figure 1.1: Two-body configurations

that could stabilize *any* configuration of rigid bodies, even without any tensile members present. The string connectivity actually in place for the operational design (where external forces may be present) might be quite different than the string connectivity that would be required to stabilize in the absence of external forces. Thus the definition of a tensegrity configuration or rigid bodies depends only on the *existence* of a set of strings that could stabilize the configuration in the *absence* of external forces.

If the configuration of rigid bodies cannot be stabilized with any set of strings (tensile members), then the configuration is not a *tensegrity configuration.* Notice that the configuration in Figure 1.1(a) is not a tensegrity configuration, while Figure 1.1(b) is a tensegrity configuration. If the system were actually built as in Figure 1.1(b), without strings, we would still classify it as a tensegrity configuration. Figure 1.1(c) is a stable embodiment of the same tensegrity configuration, hence a *tensegrity system.*

> A *tensegrity system* is composed of any given set of strings connected to a *tensegrity configuration* of rigid bodies.

From these definitions it is clear that a *tensegrity system* can be stable or unstable. By definition there exists a stabilizing connectivity for the given *tensegrity configuration*, but the actual implementation may not have adequate connectivity to stabilize. We refer to Figure 1.1(b) as a *tensegrity configuration*, and an *unstable tensegrity system.* We refer to Figure 1.1(c) as a *stable tensegrity system.*

There are many definitions of stability. One must talk about stability of a specific solution of a specific system composed of a rigid body *configuration*, a string *connectivity* and a specified set of *external forces.* There might be many equilibrium solutions, but some are stable equilibria and some are unstable equilibria. We examine only the stability of the null solution, i.e., the solution close to the specified configuration.

> The null solution of a tensegrity system (specified by a given *tensegrity configuration*, a given *string connectivity*, and a given

> set of *external forces*) is a stable equilibrium if the structure returns to the original given configuration after the application of arbitrarily small perturbations anywhere within the configuration.

Clearly tensegrity falls within the class of *prestressable* structures, so why use the word tensegrity at all? The answer is that so much has been written about tensegrity that this special class of prestressable problems deserves its proper place in the mathematical set of tools that can be specialized for this class. It turns out that this special class has important engineering applications, and this is our main motivation.

The above definitions allow significant advance in the understanding of tensegrity structure design. If the system can be prestressed to get the *same* configuration as if external loads were applied, then this provides a rule to choose the magnitude of the prestress in practical designs. That is, this feature (*prestressable to the same configuration*) allows the prestress to be chosen to yield the same configuration in the factory or lab (where the external forces are absent), as will be achieved in the operational environment where the external forces are present. In other words the tensegrity can be designed to have the same configuration loaded and unloaded. This will simplify the mathematics of design, because it sets the rule for choosing the prestress.

To distinguish between various types of systems that fit this general tensegrity definition, we offer one more distinction.

> A tensegrity configuration that has *no contacts* between its rigid bodies is a *class* 1 *tensegrity* system, and a tensegrity system with as many as *k rigid bodies in contact* is a *class k tensegrity* system.

One may wonder why precise definitions are needed after more than 50 years of the existence of the "tensegrity" concept. The answer is mathematical. We will show that these definitions above allow the simplest mathematics, and this simplicity allows deep penetration into the solution of optimization problems. Since the rigid bodies are not compliant, there are no elastic bodies in bending, only elastic axially loaded strings. Thus bending of material is not present nor required in this model. This is a great advantage in accuracy of our models of the system statics and dynamics, since bending models are quite inaccurate, compared to models of axially loaded material. Hence, generally speaking, the force that will buckle a bar is not as accurately predicted as the force that will yield the material.

Note that indeed Buckminister Fuller's definition carries the *spirit* of all these formal requirements, where stability (integrity) is achieved by tensile forces. Kenneth Snelson was perhaps the first to build a three-dimensional tensegrity structure from two rigid bodies. In the photo of Figure 1.2, note that the two wooden "X" pieces are each rigid bodies, and that these two rigid bodies are not in contact. Such a configuration of rigid bodies is a tensegrity

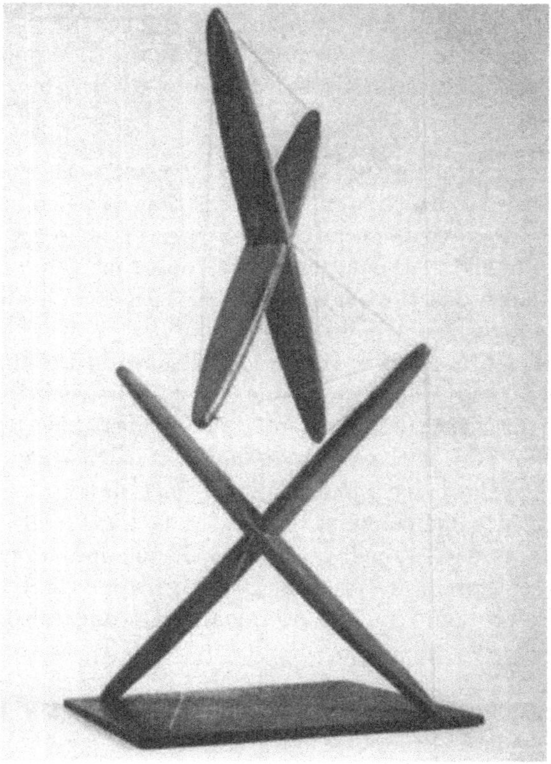

Figure 1.2: Snelson's X-piece, 1948. A two rigid body three-dimensional stable structure

configuration since the configuration can be stabilized by tensile members connecting the rigid bodies. This art piece was built by Kenneth Snelson in 1948. If some of the wires in the X-piece were missing and the structure collapsed, it would still be called a tensegrity system, albeit an unstable one, because a stabilizing string *exists* for the given rigid body configuration.

Figure 1.2 shows the first class 1 tensegrity made from two three-dimensional rigid bodies. Figure 1.3 shows the first known construction [Uit22] of a three-dimensional tensegrity using three bars and the minimal number of strings (9), by Ioganson, 1920–1921. We label these *class* 1 *tensegrity* structures.

Figure 1.4 illustrates the simplest example for a tensegrity system: one rigid body and one tensile member. A crude approximation of a violin might look like this. Figure 1.5 shows two other examples. On the left are two rigid bodies connected at a frictionless ball joint, with two strings to stabilize the configuration. On the right, three rigid bodies are stabilized by a minimum of three strings.

Figure 1.3: A three bar, nine string three-dimensional stable structure [Uit22]

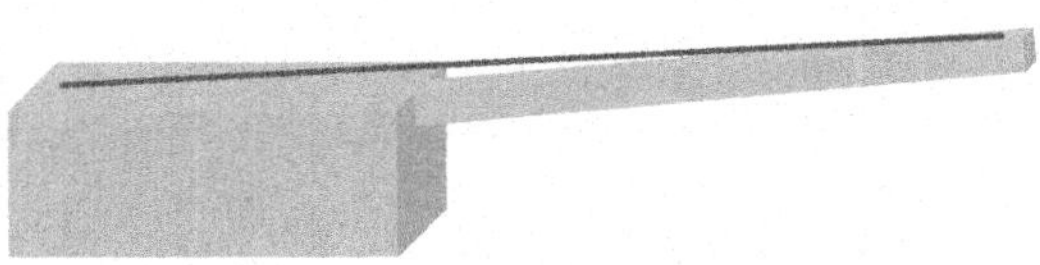

Figure 1.4: A single rigid body, a single string class 1 tensegrity system

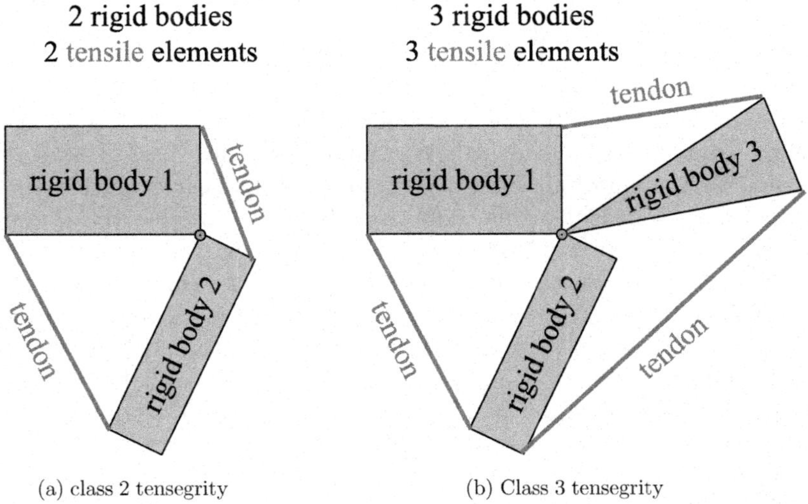

(a) class 2 tensegrity

(b) Class 3 tensegrity

Figure 1.5: A class 2 and a class 3 tensegrity system

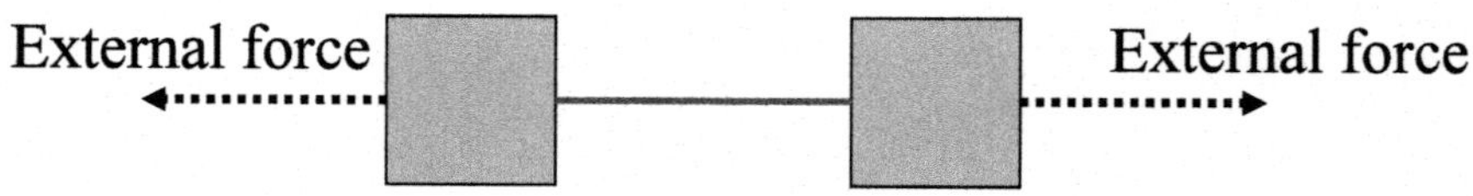

Figure 1.6: Two rigid bodies and a single string that do not constitute a *tensegrity configuration*

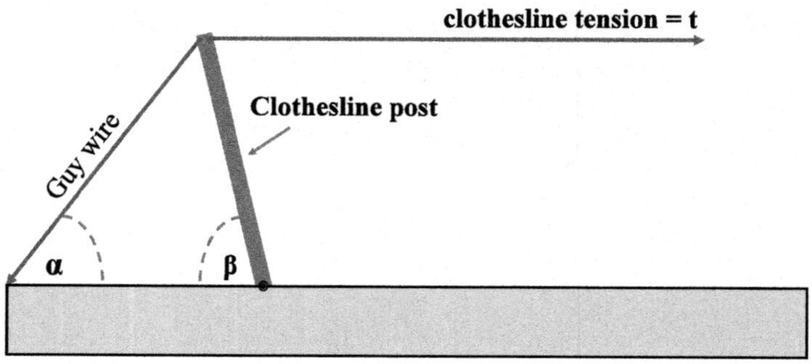

Figure 1.7: A class 2 tensegrity model of a clothesline, with two rigid bodies (pole and earth) and two strings. For $\alpha = 45°$ then, for minimal mass, $\beta = 45°$ if failure is due to yielding or $\beta = 80°$ if failure is due to buckling of the pole

Figure 1.6 is composed of two masses and one string. This configuration is not a *tensegrity configuration* because the system is not prestressable in the *absence* of the external forces.

We now preview the type of optimal design that will be studied in this book with a simple example. To erect a clothesline in the backyard (well, times have changed with the availability of clothes dryers), typically one mounts a pole perpendicular to the ground. However, if one wishes to use the smallest amount of material to construct the clothesline, one can find the minimal mass solution to this problem as follows. Consider the clothesline in Figure 1.7, where the pole is fixed by a ball joint at the ground, and the clothesline wire is assigned a specified tension t. Only half the system is shown in the sketch. One might consider this a class 2 tensegrity system with two rigid bodies (the earth and the bar of the clothesline), similar to the class 2 example of Figure 1.5. Note that the earth as part of our system in the clothesline example can be interpreted as in Figure 1.4, where the violin rigid body plays the role of the earth-like boundary.

The mathematics of Chapter 3 yields structures with minimal mass. For the clothesline example, if *buckling* is the mode of failure, then the pole angle β and the guy wire angle α must satisfy $4\tan^2\beta + 5\tan\alpha\tan\beta = 1$. If $\alpha = 45°$, this would fix the pole angle at $\beta = 80°$. On the other hand, in the unlikely situation that material *yielding* is the mode of failure of the pole (instead of buckling), then the math of Chapter 3 would require $\alpha = \beta = 45°$.

Both of these results remain independent of material choice for the pole or wire, and independent of the wire tension. The pipe or pole diameter will be fixed by the buckling calculations, involving the choice of material and the magnitude of the force (tension in the clothesline), but the angle of the pole (80°) is optimal independently of the choice of pole material, and independent of wire tension, as well. One benefit of our optimization for tensegrity systems is that quite often the optimal geometry of the configuration (optimal topology) can be determined independently of material choices, and independently of the magnitude of external loads.

1.1 Tensegrity in Nature

It may seem strange to begin an engineering book with discussions of art, beauty, and biology. Yet to serve important mechanical, electrical, and other functions, we observe that nature uses simple non-toxic (biodegradable) materials in a sophisticated (and many times beautiful) architecture. D'Arcy W. Thompson says "The Book of Nature may indeed be written in the characters of geometry." Indeed, there is much to learn from nature. Before proceeding, however, we remind the reader that mathematics cannot model any physical system exactly. We can only approximate nature with simplistic models. We say that the mathematical model is an *idealization* of the truth. Of course, there is obvious benefit in choosing a structural paradigm that fits the system as close as possible, in the sense that useful features are captured while the required mathematics is kept as simple as possible.

The bones and tendons of animals and man are connected in a manner that allows easy control of movement. Obviously, these structures evolved for control functions, where the bones provide compressive load-carrying capacity and the tendons provide the stabilizing tensions required in a given configuration. Figure 1.8 suggests how tendons are connected to bony parts to control them. Note that the tendons (with muscle actuators) connect the humerus bone of the arm with the ulna and radial bones of the forearm. These three bones intersect at the elbow, hence we classify this as a class 3 tensegrity joint. Note also that the bones (especially the cupula) are not rods. Indeed, the shape of the rigid bodies each have a purpose. The second sketch in this figure suggests that a class 2 tensegrity approximation of the toe control system might be reasonable.

Figure 1.9 shows the locomotion kinematics of a cat's hind legs. The red and blue plots are forces in the respective tendons that are designated as flexors (red) and extensors (blue). This figure is from the work of Orjan Ekeberg [EP05, EP06, PEB06, HE08].

A class 1 tensegrity idealization seems to fit very well the molecular structure of the spider fiber. The molecular structure of nature's strongest fiber,

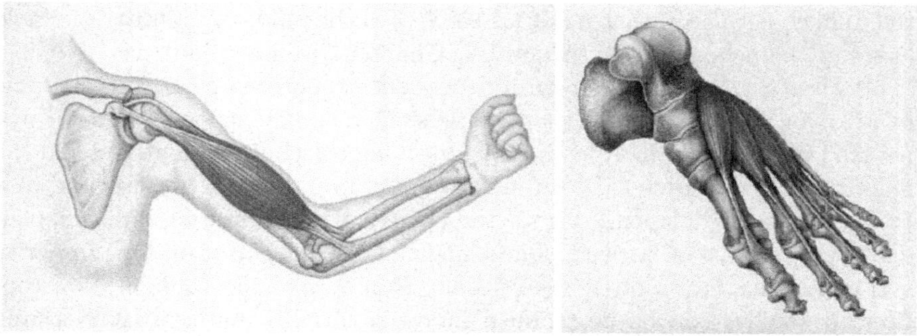

(a) class 2 shoulder joint and class 3 elbow joint (b) class 2 toe joints

Figure 1.8: The elbow can be thought of as a class 3 tensegrity system, the shoulder as a class 2 tensegrity system, and the foot as a class 2 tensegrity system

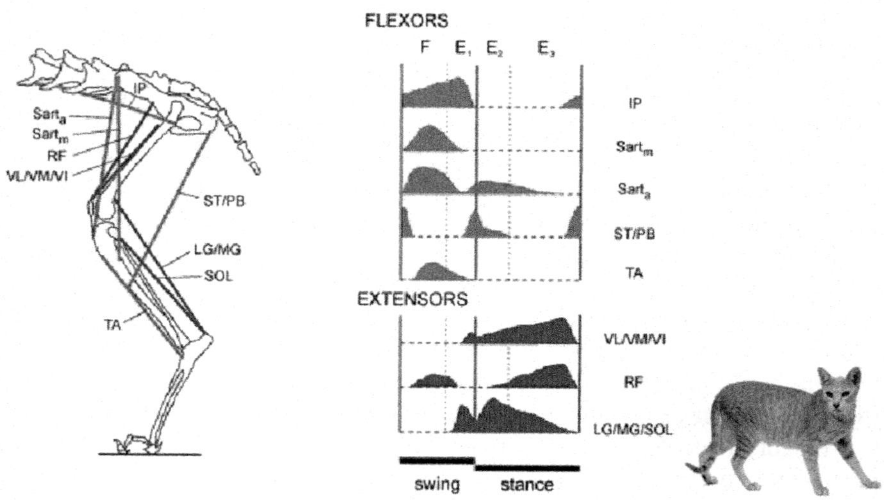

Figure 1.9: The flexor tendons (*red*) and the extensor tendons (*blue*) of a cat's hind legs. The plot shows the time profile of the forces in each tendon during a walk

the dragline silk of a Nephila Clavipes (a spider in Figure 1.10, commonly known as the golden orb weaver) has a tensegrity architecture. The Cornell researcher Lynn Jelinski [SMJ96] and DuPont's Y. Termonia [Ter94] explain that the spider silk is a complex-folded protein composed primarily of two amino acids, glycine and alanine. The alanines are aligned in two ways to form (1) *rectangular plates* (molecular plates from tiny crystals), called β-pleated sheets in Figure 1.10, and (2) *amorphous strands* making a tensile network of material that can take up the strain. The rectangular plates provide the rigid bodies in our tensegrity definition, and the amorphous strands

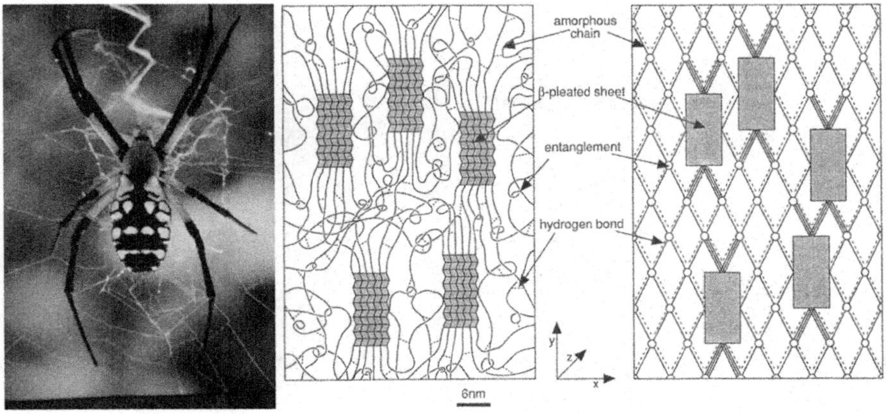

Figure 1.10: A class 1 tensegrity model of the molecular structure of the spider fiber. The rigid bodies are the β-pleated sheets, and the tensile members are the amorphous strands that connect to the β-pleated sheets

form the tensile members of our tensegrity definition. Since the β-pleated sheets are not in contact, the material of the spider fiber is stabilized by tensile members. As in the clothesline example, if we count the rigid earth, which establishes immovable boundaries for the spider fiber extremal attachment points, as part of our *tensegrity configuration* of rigid bodies, then the spider fiber forms a class 1 tensegrity system, idealized somewhat to fit very well the molecular structure of the spider fiber.

The membrane of red blood cells have remarkable static and dynamic behavior. Attached to the under-side of the red blood cell membrane (the lipid bilayer) are 33,000 units that have convenient tensegrity models. Refer to the cartoon in Figure 1.11, where the protofilament (the junctional complex) is rigid compared to the rest of the components, so it is modeled as rigid. The elastic components are the spectrin (tendons). Biologists have determined the exact binding sites where the spectrin attach to the junctional complex (rigid rod). In [VSBS05] the three-dimensional nanomechanics were derived using the tensegrity model of an erythrocyte junctional complex in equibiaxial and anisotropic deformations. The complete red blood cell has 33,000 units of these junctional complexes underneath the membrane (lipid bilayer) of each red blood cell. The paper [VSBS05] describes the static and dynamic behavior of an individual junctional complex, in the presence of a variety of initial conditions that might represent an unhealthy state of the cell. By modeling the protofilament as a rigid body, and by modeling the spectrin as the tensile connectors between the lipid bilayer (membrane) and the protofilament at the precisely known attachment points, the equilibrium position of the protofilament was determined (using lab data for spectrin stiffness (due to folding it is very nonlinear), and the geometry of the tensile

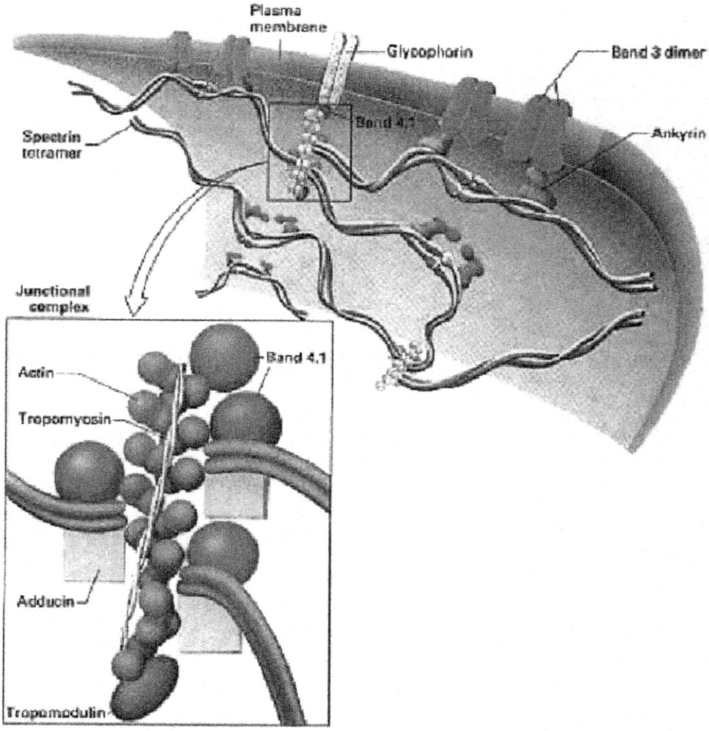

Figure 1.11: The network of junctional complexes underneath the red blood cell membrane. The protofilament (33,000 in each red blood cell) is the rigid body and the spectrin dimer are the tensile members. Each spectrin is stapled to the lipid bilayer

network from atomic force microscopes). Some of these results are shown in Figure 1.12. The paper [dOVV+09] continues this work modeling networks of junctional complex. Information that can be obtained from the tensegrity model include the equilibrium position of the protofilament, relative to the plane of the membrane and also relative to adjacent units in networks. Values obtained by simulation seem to be in very good agreement with experimental observations [VSBS05, dOVV+09].

Other tensegrity models of other cell cytoskeletons whose mechanical behavior in living cells is consistent with a tensegrity model have been proposed for many years by the biologist Don Ingber at the Harvard Medical School [Ing98], and others [WOI99, WNS+01, SSI04, SMJ96, Pea90, Kob76, CLO+02, CS97]. In the 1998 issue of *Scientific American* Don Ingber claims that "tensegrity is the architecture of life."

In spite of our progress in engineered structures and materials, man-made designs still cannot approach the efficiency (power per unit mass) of biological

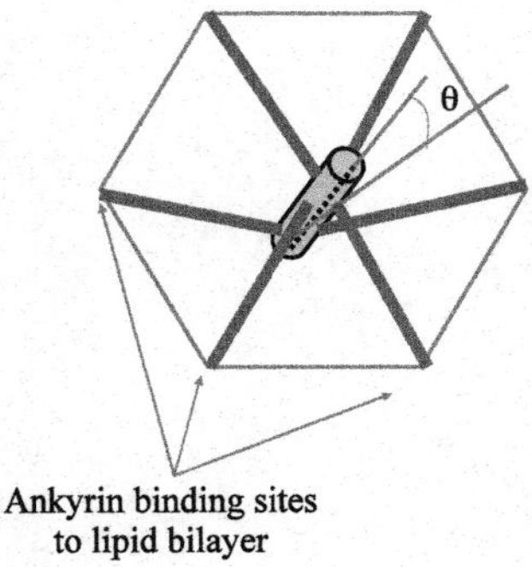

(a) Single Unit

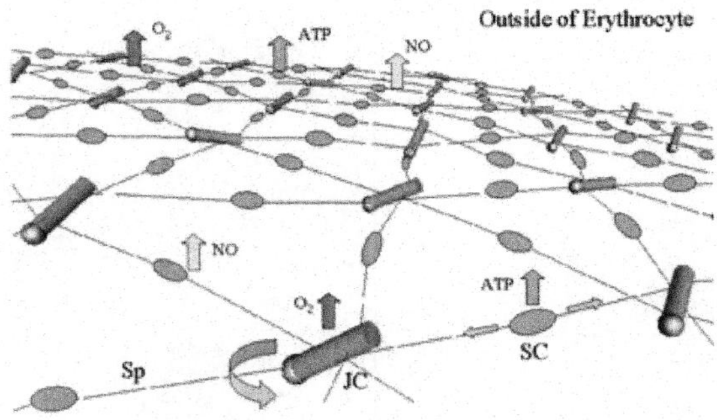

(b) Small Network

Figure 1.12: (a) This figure depicts the steady-state results for a single junctional complex within the network of 33,000 junctional complexes underneath the red blood cell membrane. Each of the six binding sites where the strings connect to the rigid body are known and can be found in [VSBS05]. In the tensegrity model, the actin protofilament (rod) has a radius of 4.5 nm and a length of 37 nm. The equilibrium pitch angle obtained was $\theta = 17.8°$. The equilibrium tension in each Sp dimer (cable) is given in [VSBS05]. (b) This figure shows the steady-state results for networks of junctional complex superimposed with some extra arrows indicating how the mechanical equilibrium might relate with certain physiologic responses [dOVV$^+$09]

systems. Nature seems to produce outstanding *systems* designs, while man concentrates on *components* of a system but fails to have rules on how to put components together to make an efficient system with a prescribed set of functional capabilities. Humans make large factories to produce small products, whereas nature can produce products larger than the factory.

Integration of mechanical and electrical functions could be accomplished by optimizing the material topology and the material choice. Although this book focuses only on the optimization of mechanical properties, books will undoubtedly follow that show how to optimize topology to get special acoustic properties or electrical properties. The heart muscle is an example of a structural system with combined mechanical and electrical properties.

1.2 Tensegrity in Art

Given some materials, one must choose a *geometric arrangement* of the material components to make a structure. We shall loosely refer to this geometry of those material connections as the system *topology*. Our focus is mechanical or material engineering, but engineers and architects pay their respects to beauty too. Geometry has a special kind of beauty on its own. One can perceive beauty from the most complex to the least complex patterns to fill

Figure 1.13: Kenneth Snelson's "Free Ride Home", 1974, a class 1 tensegrity structure

space, from the realists who painted complex detail to the impressionists who painted less detail, to Picasso who painted even less detail, to minimal art where complex objects are perceived by the brain from a minimal number of lines or curves on the paper. The beauty of the straight line and simple geometries was admired by Plato [Philebus Plato, March 27, 355 BC].

Tensegrity components are very simple elements, often just straight lines. This type of beauty has appealed to sculptors like Kenneth Snelson [Sne65] and architects like Buckminister Fuller [Ful59]. Figure 1.13 is a class 1 tensegrity sculpture built in 1974 by Kenneth Snelson, New York. Figure 1.14 shows two more Snelson creations.

The intersection of art and science, form and function, comes to the fore with tensegrity. The artist Kenneth Snelson himself doubted any utilitarian value of tensegrity [Sne96]. Artists lend creative and inspired concepts to structures, without the benefit of (nor the need for) analytical tools to characterize functional properties of the structure. Yet, engineers lend sophisticated analytical tools to uninspired structural concepts. At the intersection of this dilemma is the mathematical work called "fractals." Fractals are obtained by filling space (usually the plane) with an infinite number of self-similar geometric objects. Tessellation or "tiling" theory deals with filling a finite space with a finite number of similar objects in a similar topology. Mathematicians Mandelson and Stephen Wolfram [Mal00] made giant contributions to this

Figure 1.14: class 1 tensegrity towers by Kenneth Snelson

field. The artist M. C. Escher [Mac65] took tessellation to a popular art form (one author's favorite tie is an M. C. Escher creation).

Fractal mathematics and the art forms discussed above require no functional purpose, whereas biological examples of topology obviously have a functional purpose. This is not to say that scientists have not tried to connect the two. They have generated fractals and self-similar structures and compared them to the topologies found in biological material, such as in sea shell [Mal00]. See one of the fractal rules of Wolfram illustrated in Figure 1.15. The topology of natural material probably developed to solve a functional purpose for survival (be it for mechanical strength, thermal properties, or electrical, or acoustic characteristics), and not just filling space as fractals can do. We shall later discuss fractals of a special kind, where space is filled with self-similar patterns but with an added certificate of mechanical performance, such as a guaranteed strength or stiffness. Here we simply point to the kind of "Platonian" beauty of those fractals that involve only straight lines.

Tensegrity systems have been around for over 50 years as an art form [Uit22], with some architectural appeal [Pug76, Mot03, Maz03, Lal96, Gou98, Ped98, MR03], but the absence of analytical tools for convenient analysis and optimization with which to design engineering structures from tensegrity concepts has prevented such intriguing topologies from taking their rightful place among the alternatives for design and construction of engineered structures. Quoting Elisabeth Eaves from Slate.com, Jan. 9, 2004, "Art for Smart People. The mathematical sculptures of Kenneth Snelson":

> There is little doubt about the mechanical ingenuity or scientific impact of Snelson's work. But is it also art? Absolutely. His sculptures are pure outpourings of creative energy, utterly useless as objects and yet visually arresting.

Figure 1.15: A fractal developed by Stephen Wolfram [Mal00] sometimes compared with the material of sea shells

Indeed early papers on tensegrity agreed with the artist Kenneth Snelson that tensegrity is art and probably has no utilitarian value. Gunnart Tilbert and Sergio Pelegrino [Til02, TP03] point out quite correctly that tensegrity systems composed of architectures built and envisioned by Kenneth Snelson were not stiff enough for many engineering purposes. But this should not be understood as a criticism of the tensegrity concept (even though such language was used). This is simply a criticism of, and a fact about, *one* embodiment of tensegrity concepts: the one they had in mind. The type of structure examined by Tilbert is shown in Figure 1.16, which is fairly soft in bending and compression. There are, however, many tensegrity examples beyond those found by the artists. Tensegrity systems can now be made very stiff indeed, once the mathematical tools are constructed to allow a demand for high stiffness or strength.

1.3 Tensegrity in Architecture

We expect a dramatic increase in tensegrity concepts in architectural engineering, since it is now possible to demonstrate mathematically (in subsequent chapters) that tensegrity structures can be designed to efficiently take tension, compression, or bending. As material costs increase, it is reason-

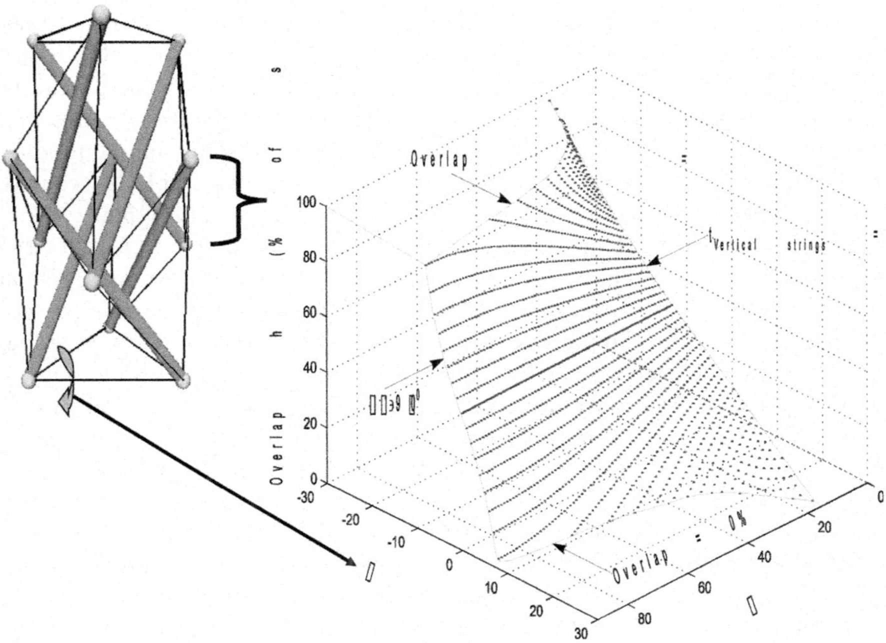

Figure 1.16: The equilibrium surface for a class 1 tensegrity tower, using *regular minimal tensegrity prisms*

able that methods that make more efficient use of material will become more acceptable, even if that requires rewriting some outdated building codes.

But the advantages of tensegrity in architecture goes beyond mass efficiency. Nature uses tensegrity whenever a large controlled change in configuration of the structure is required (such as in animals and cells). Controlled tensegrity can alter the amount of solar energy a building interior receives, to make buildings more energy efficient, by making the building responsive to external events, such as earthquakes, winds, and thermal loads [dS03, dS06a, dS06b, dS06c, Maz03].

There are important reasons to pursue responsive architecture [dS06b]. One reason why a building should be designed to change its configuration is to make better use of the natural environmental disturbances acting on the building. That is, one should use these disturbances to advantage in case they can be beneficial (such as increased solar lighting or heating), and to reduce the damage caused by these disturbances in case they cannot be beneficial (e.g., reducing the impact of an earthquake or high winds from a hurricane).

Beneficial disturbances: From time to time the mission for a given space may change, requiring different lighting or heating conditions. In this case buildings will modify the shape of the space to suit the user, within limits.

This has already happened in a limited extent. The superdomes can adjust the amount of light in the stadium by opening or closing the roof. "Sun roofs" in homes will take on a new meaning, adjusting the entire roof exposure to the living room or the patio, or the kitchen. Walls separating patios and living rooms can be automatically moved or closed, depending upon humidity, sunlight, rain, etc. Windows and roofs will both be adjustable to regulate heat and airflow, to reduce the cost of heating and air-conditioning.

Detrimental disturbances: In order to survive violent movements of the ground and or high winds, the building itself must move, just as a person might bend his knees or stiffen (or soften) his body and joints, or lean into the wind to stand in the presence of ground motion or high winds. Designing a building to survive violent movement of the foundation will lead architects and contractors to better buildings that will reduce the burdens of insurance companies that must cover the losses in an earthquake and reduce the number of casualties in a natural disaster. In 1979 Japan built the first controlled building, where a *momentum exchange controller* was installed to move a large mass on rails on the eleventh floor of the building in response to earthquake ground motion. The building moves less by arranging the large mass to move in opposition to the earthquake motion. In the future, a more efficient controller will be developed that does not add 10% of the mass of the building on the eleventh floor to counteract the motion. Instead, tendon-controlled cables will be installed to control the motion more efficiently, and more precisely, and more reliably. These are tensegrity applications just waiting to be developed.

New construction techniques: Incorporating motion control capability into the structure is not as simple as adding a new controlled actuator technology to an existing structure design method. Both the design of the structure and the controls have to be coordinated to efficiently allow the required movement. This will create new techniques for construction. There exist companies even now dedicated to this task of developing new design and construction techniques to integrate the structure with the movement "muscles." The most notable progress so far is deployable structures that have been designed to be transported to a construction site in a stowed compact package, and then deployed with only the use of winch power. Such deployable houses or field hospitals can be delivered (dropped from cargo planes or from helicopters) to a disaster area or provide housing for third world countries. In the near future such deployable housing concepts will make permanent homes and other buildings.

A simple tensegrity shelter is shown in Figure 1.17, which is a class 2 tensegrity structure, due to the bar-to-bar connections around the periphery. Such structures could be stowed in a small package and dropped on a disaster site, and deploy with winch power (manual or electrical) to function as a field hospital or temporary housing.

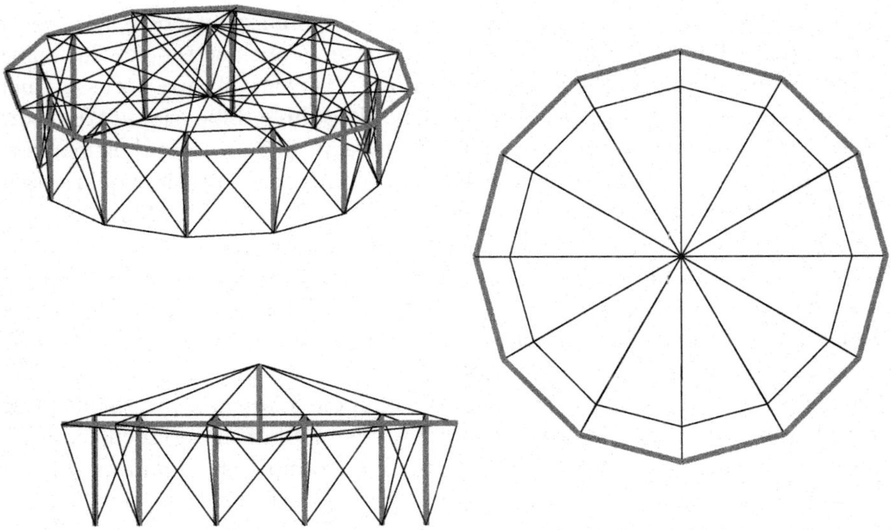

Figure 1.17: A class 2 tensegrity shelter

Even though a tensegrity structure might look complex, the appropriate efficiencies can be demonstrated by sound analytical tools and an algorithm for construction technique. In fact, design costs, building costs, and reliability are not necessarily measured by complexity but depend more upon the availability of analytical tools to reduce such designs to reliable algorithms and construction techniques.

1.4 Tensegrity in Engineering and Science

In art, it is sufficient for the tensegrity to have beauty. But, the engineering world should not embrace the concept unless some fundamental questions are answered, such as "What precise mathematical problem is solved by a tensegrity system?" This book intends to answer such questions. Any loaded structure might have parts in compression, parts in tension, and other parts in torsion. So, if one could know the optimal arrangement of material to take a tensile load, or a compressive load, or a bending load, then one would have a great start toward designing an efficient structure for the whole system. We will show that indeed, an efficient, and sometimes optimal, arrangement of material for each of these cases (tensile, compressive, or bending loads) is a tensegrity structure.

Early man-made structures employed orthogonal structural elements, such as the architecture of Stonehenge, or the ailerons, and spars in the classical wing, or beams and columns in a building. While such orthogonal elements are simple to manufacture, there is ample evidence (see the biological examples above) that a topology composed of orthogonal elements is not necessarily mass efficient for the functional purpose. Consider the revolutionary improvements in mass efficiency enabled by the invention of the truss, which departs from orthogonal material topology. The three-dimensional truss concept can be made even more efficient yet, by exposing structural elements to only unidirectional loads. Such features not only simplify the equations of motion, but the resulting models will be much more accurate than models of bodies that are subject to bending moments. That is, if the internal stresses in the rigid bodies have a *specific predetermined direction*, then that feature assures more accurate dynamic models. In addition to modeling advantages, some mass efficiency can be obtained if the load directions within each body can be pre-specified. If each bar can be loaded in only one direction (axially), the *material* choices can be specialized to handle loads in a pre-specified direction with much less mass than would be required of material choices that must take loads in a variety of directions. For example, sand and mortar are very good materials in compression but are not used in situations where the dynamic loads can reasonably be expected to reverse directions, as in severe earthquakes. The loads in a structure can be made unidirectional by applying sufficient prestress [Pel90, YP96]. For example, a prestressed concrete beam avoids tensile loads in the concrete. The goal in tensegrity systems is to assure that the members are unidirectionally loaded, so that no member is required to serve both compressive and tensile functions.

The largest body of scientific literature on tensegrity structures deals with the form-finding problem, where one searches for a configuration, and a set of admissible member forces, to satisfy equilibrium conditions [Emm59, Emm96, CP91, Cal78, KKAM99, Leo88, Max64, RW81, Tar80, Whi84, OW97, MNJ86, VM99, PC86, MN01a, ZMM06, MSG05, MSG06, SS97a, Ske05, KW02b, KW02a]. One of the first attempt to characterize dynamics of class 1 tensegrity systems, a network of rods and strings, appears in [Pin05, SPM01, Ske05, SCS02] some 50 years after the invention of the tensegrity concept. In [Pin05] the vector equations are very complex indeed. Chapter 5 will provide a much more convenient analytical tool to describe both the statics and the dynamics of tensegrity systems. Other approaches to dynamics appear in [Lan83, HP79, dO06, CW92, CW96, CB98, ASKD03, Mur01, MN01b, OW01b, OW01a, OW00].

The essence of this book is to seek a methodology to use simplistic elements (nothing new here) to fill the allowed space (new here), with guaranteed mechanical properties (a lot new here). These include static as well as dynamic properties with and without the use of active control action. Some discussion on the principles and simple examples of engineering applications of tensegrity follow.

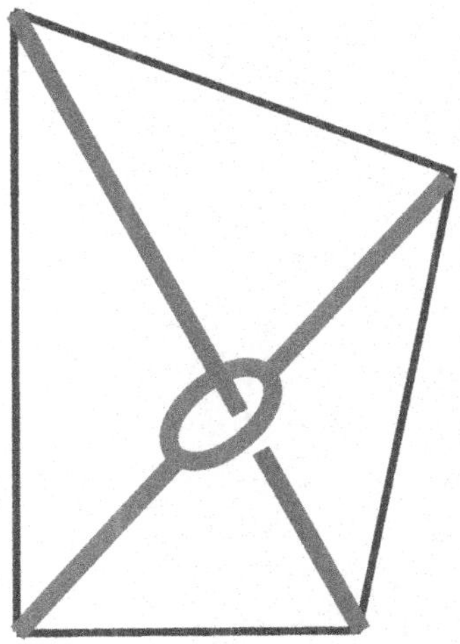

Figure 1.18: The simplest nontrivial tensegrity structure with bars and strings

1.4.1 Fundamentals of Tensegrity Structures

Basic building blocks

The simplest tensegrity is a single rigid body and a string, as in Figure 1.4. In the case of bars and strings, this is simply a prestressed bar. The next simplest, and first nontrivial tensegrity structure with bars and strings, is two bars and four strings, as in Figure 1.18. The next simplest is a fundamental three-dimensional unit we will call a *Tensegrity Prism*, which we discuss in more detail next.

A *Tensegrity Prism* is composed of any stable three-dimensional unit using p bars, with an p-sided polygon of strings on the "top" and an p-sided polygon of strings also on the "bottom" of the unit. Apparently, Ioganson [Uit22, Gou98], was first to build such a unit for $p = 3$, but Snelson made it popular in an art form. Several variations of *tensegrity prisms* will be useful. The photo of a *regular minimal tensegrity prism* appears in Figure 1.19 for $p = 3$, where the top view is photographed in Figure 1.20. The word *minimal* refers to the use of the smallest number of strings ($3p$) to stabilize, and the word *regular* refers to the requirement that the top and bottom polygons are parallel and equilateral (but the top and bottom polygons do not necessarily have the same radius, as was the case in Figures 1.19, and 1.20).

The *regular minimal tensegrity prism* in Figure 1.20 contains two equilateral triangles, one formed by three strings at the top and the other formed by three strings at the bottom. These two triangles are parallel but may not have the same radius (even though they do in this figure). The remaining three strings connect the vertices of the top triangle to the vertices of the

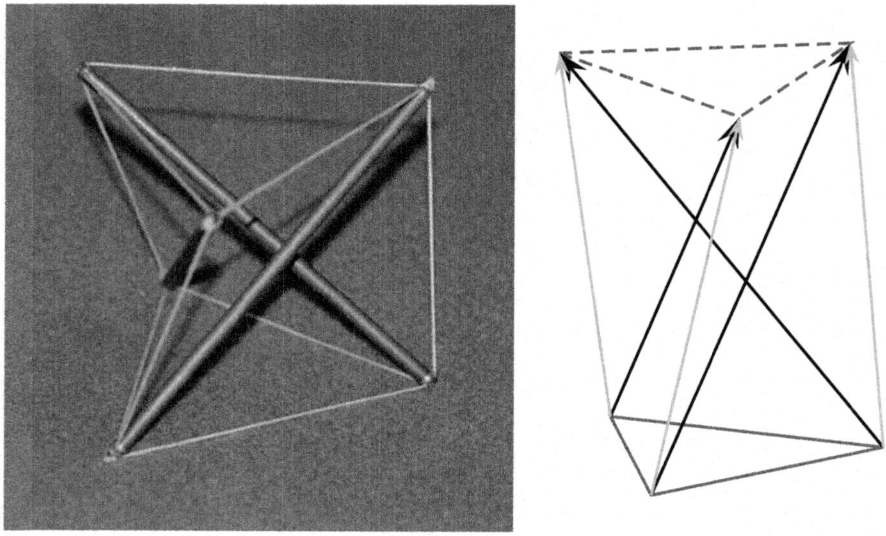

Figure 1.19: Perspective view and diagram of a *regular minimal tensegrity prism* with same top and bottom radius

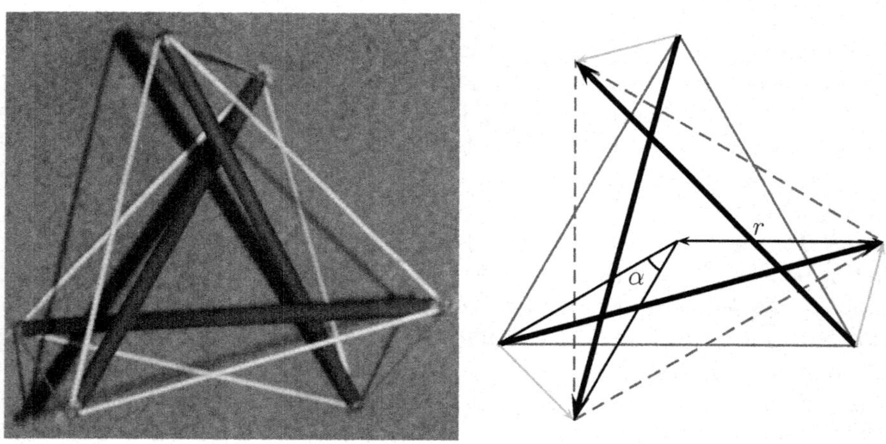

Figure 1.20: Top view of A *regular minimal tensegrity prism* for $p = 3$. The only equilibrium of the unloaded prism is at 30° as shown. Red solid lines form the bottom equilateral triangle of strings, blue dashed lines form the top equilateral triangle, green solid lines form the strings connecting top to bottom triangles, and the black lines denote the bars which also connect top and bottom triangles. For three-bar prisms the unforced equilibrium is $\alpha = 30°$, and the equilibrium in the loaded case can approach $\alpha = 90°$

bottom triangle. The bars (brass rods in the photo) also connect certain vertices of the top triangle to the vertices of the bottom triangle. In Chapter 3 we will show that in the unforced case, the only stable equilibrium of the three bar ($p = 3$) *minimal regular tensegrity prism* requires the twist angle between the top and the bottom triangles to be precisely 30°, and this fact remains true regardless of the radii of either top or bottom triangles. Such conclusions follow from the general result below, for any choice of p (as per Section 3.5).

For any *regular minimal tensegrity prism* composed of p bars, let an external force of magnitude f/p be applied at each of the $2p$ nodes perpendicular to the top and bottom of the prism (formed by p-polygonal faces), so as to compress the prism to a smaller height. The equal forces will keep the prism *regular.* Summing forces acting on each bar one can derive the equilibrium condition. In the absence of any external forces ($f = 0$) the prism equilibrium is uniquely $\alpha = \pi/2 - \pi/p$, as mentioned above. In the presence of the type of external forces described above, the prism twist can be any value in the range $\pi/2 - \pi/p \leq \alpha \leq \pi - 2\pi/p$, where the upper bound corresponds to an intersection of the p bars at the middle of the structure. Bars would actually interfere at angle somewhat less than $\alpha = \pi - 2\pi/p$, due to bar thickness.

Primal and dual structures

We now introduce a concept that applies to any configuration of bars and strings. Let a given configuration of bars and strings be called the *primal* system in this discussion. Now define another system, called the *dual,* which has the same geometric appearance as the *primal,* but the strings have been replaced by bars, and similarly the bars have been replaced by strings. Some examples of *primals* and *duals* follow.

Imagine just three bars connected end to end in a plane. This is the *primal* system in Figure 1.21(a). Now imagine the *dual* of this system, which is a three-string system without any bars. Obviously neither the primal nor dual qualify as tensegrity configurations. The *primal* requires no string to stabilize this configuration (the three-bar system is already stable and indeed rigid), so the primal is not a tensegrity because of the absence of strings. The *dual* cannot be stabilized because no bars are present, so this is not a tensegrity system because it is not stabilizable.

Consider the tensegrity *primals* and *duals* illustrated in Figure 1.21, where the strings are red thin lines and the bars are gray thick lines. In Figure 1.21(b) the configuration of the two-bar system is in a stable equilibrium if, and only if, the bars cross each other. That is, the configuration cannot be stabilized by any string placements if both ends of one bar lie on the same side of the other bar. So, except for this condition, the left hand figure (the *primal*) is a tensegrity configuration in both two-dimensional and 3-dimensional space. The dual of this system is a four-bar system. This configuration, unlike the primal system, cannot be stabilized in three-dimensional space but

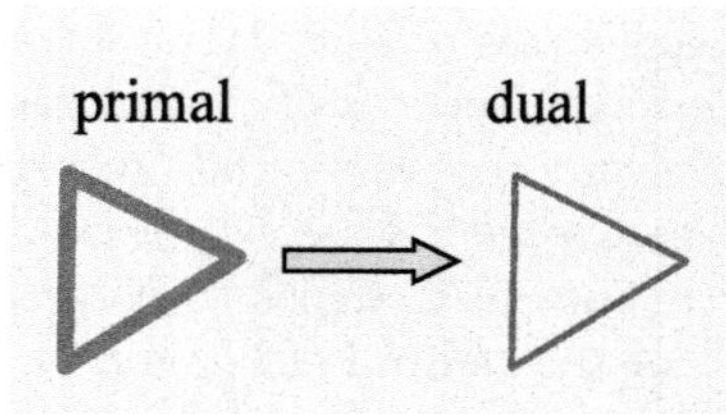

(a) Primal and dual (not tensegrity)

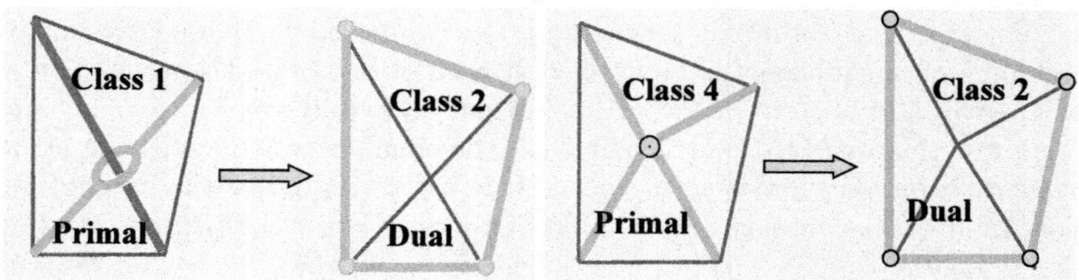

(b) Stable primal and 3D unstable dual

(c) 3D unstable primal and dual

Figure 1.21: Any system of bars and strings (*primal*) has a *dual* obtained by replacing bars by strings and vice versa

can be stabilized in two-dimensional space. Note from these examples that the *dual* of a tensegrity system may or may not be a tensegrity system.

In Figure 1.21(c) the four rigid bars Figure 1.21 form a class 4 tensegrity in the plane. Of course this configuration cannot be stabilized in three-dimensional space, since the third dimension of the configuration is not stabilizable. Neither is the configuration of the *dual*. Any tension applied in one string would immediately collapse the system to fold about the axis of the other string, moving the system out of the plane. Both *primal* and *dual* are tensegrities defined in a two-dimensional space.

For $p = 3$, the *dual* of the *regular minimal tensegrity prism* is the class 3 prism that results when the strings and bars reverse roles in the right-most sketch in Figure 1.20, where strings are now black (three of them), and the red, green, and blue lines are all bars (nine of them). This is a class 3 tensegrity structure, whereas the primal was a class 1 tensegrity structure.

Minimum mass

For p bars, with a total crushing force f (each node receives force of magnitude f/p), the minimal mass for each bar in a *regular minimal tensegrity prism* is given by m, where,

$$m^2 = \frac{4f\rho^2 \sin\alpha[h^2 + 2r^2(1-\cos(\alpha + 2\pi/n))]^{5/2}}{\pi n h E(\sin\alpha - \sin(\alpha + 2\pi/n))}, \tag{1.1}$$

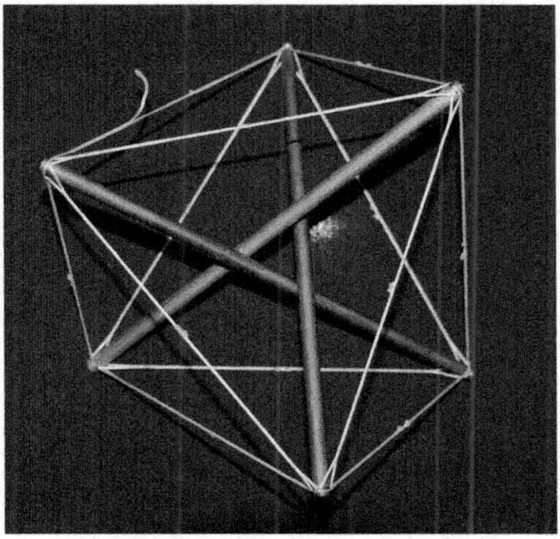

Figure 1.22: A stiff regular non-minimal tensegrity prism. The minimal number of strings that can stabilize is nine. This structure has 12 strings

where E is Young's modulus, ρ is mass density, r is the outer radius of the p-polygon, h is the height of the prism, f is the total crushing load (distributed evenly over the nodes), and α is the twist of the top p-polygon with respect to the bottom p-polygon. The length ℓ of a bar in terms of the height h and radius r, and α is

$$\ell^2 = h^2 + 2r^2(1 - \cos(\alpha + 2\pi/n)). \tag{1.2}$$

See early discussions on this problem in [QKAM03, Han92a, Han92b, Han94].

Non-minimal structures

The point has already been made that tensegrity structures can be stiffened by adding strings at special places. Consider that the *regular minimal tensegrity prism* has a soft mode (see Chapter 2), but by adding three extra strings as in Figure 1.22, the prism is stiff. One has a choice of using building blocks made of *regular minimal tensegrity prisms*, which are soft, or *regular tensegrity prisms*, where the extra strings can be strategically located to increase stiffness and remove any soft modes. At the date of this writing, we have only seen *minimal tensegrity prisms*, as the building blocks of tensegrity structures both in art and in architectural studies, and this may explain the reluctance of engineers to adopt tensegrity designs for engineering.

Changing shape and stiffness, deployment

Tensegrity is a structural paradigm that has the very unusual property that the stiffness can be changed without changing external shape, and conversely

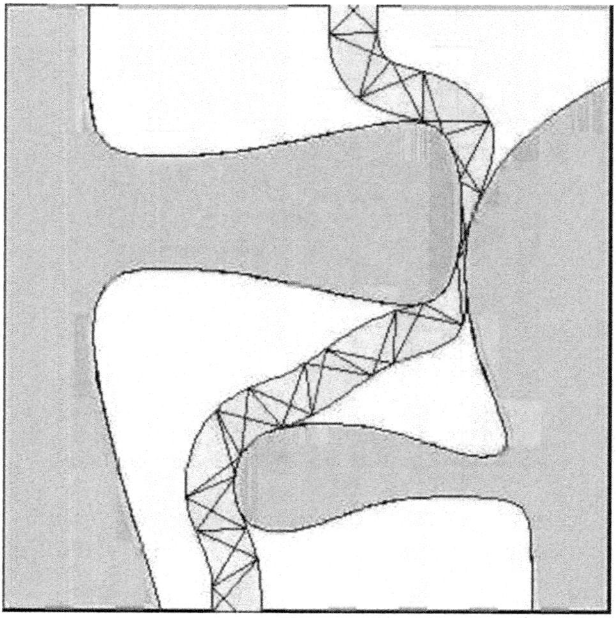

Figure 1.23: A tensegrity worm crawling through crevices. From the thesis of Jack Aldrich [ASKD03]

the shape can be changed without changing stiffness. This is due to the fact that internal changes in geometry can change stiffness, as measured externally. Likewise, special changes in the internal configuration can change stiffness without changing external shape. Recall that the *configuration* defined earlier includes the position of all compressive members, but the stiffness and shape might involve only certain internal or external members, so that clever choices can satisfy both kinds of constraints. This might have great impact in control of flexible and deployable tensegrity structures. Figure 1.23 illustrates the extreme shape-changing ability of tensegrity structures, where a robotic tensegrity worm crawls (which requires stiffness) while squeezing through crevices (which requires large shape changes). The mathematical details for these crawling and shape-changing controls appear in the work of Jack Aldrich [ASKD03, Ald04, AS05, AS06], who derived control laws to prevent slack strings, prevent string yielding, and prevent bar buckling.

The second nice property of tensegrity is that stiffness is largely dictated by clever choices of geometry, rather than by increasing prestress and mass. In general, stiffness depends upon both geometry and prestress, but Figures 1.24 and 1.25 illustrate that choices of geometry play the dominant role. Prestress increases the required mass directly, whereas geometry choices do not. For these and other reasons, tensegrity structures can be of very light weight compared to other design choices.

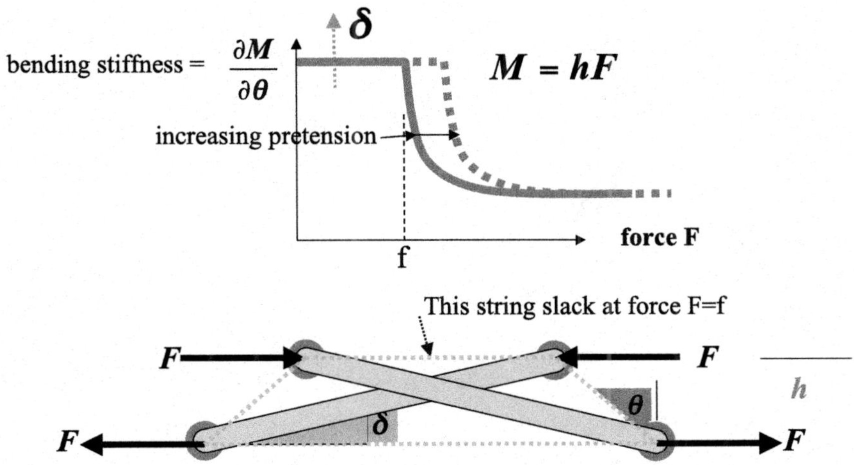

Figure 1.24: A class 1 tensegrity structure in bending. The bending stiffness is dominated by geometry, and the robustness to uncertainty in external moments is dictated by prestress. The stiffness drops when a string goes slack

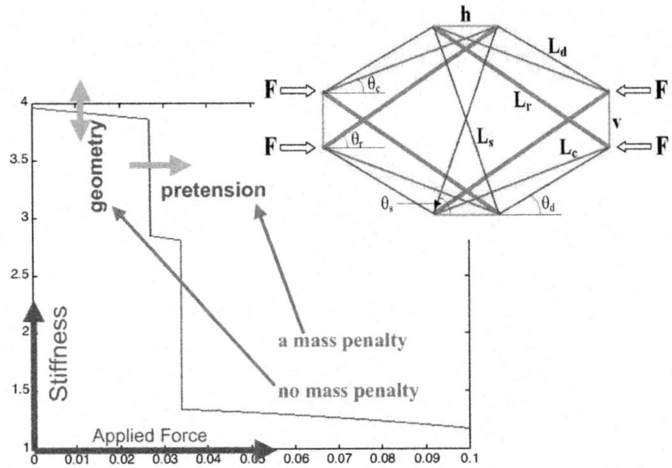

Figure 1.25: A class 1 tensegrity in compression. The compressive stiffness is dominated by geometry, and the robustness to uncertainty in external forces is dictated by prestress. The stiffness drops when a string goes slack

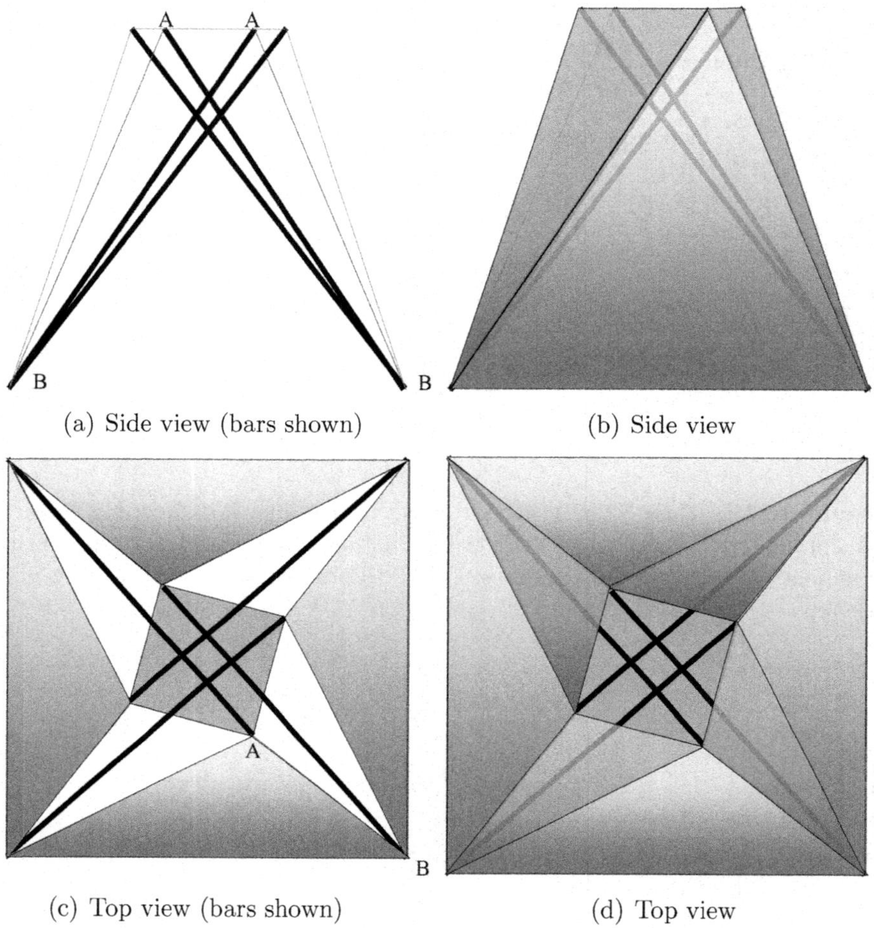

Figure 1.26: A class 1 tensegrity tent based on the *non-minimal regular tensegrity prism*

1.4.2 Temporary Shelters and Tents

The simplest example of tensegrity housing is a tent. Figure 1.26 shows a sequence of construction steps to make a tent with four bars. This is not a *minimal tensegrity prism*, because of the extra strings along lines A–B. The simplest tent would require three bars (poles) and nine strings, which would be the simple *regular tensegrity prism* configuration of Figure 1.20. In our example, however, we depart from the three-bar *regular minimal tensegrity prism* which is soft in a certain rotational direction and gives little interior space. A four-bar regular minimal tensegrity prism would give more interior space, but would also be soft, and perhaps perform poorly in wind storms. Instead, we choose a *regular tensegrity prism* which is not minimal, but has four bars and 16 strings, as in Figure 1.26. The extra strings that make it non-minimal are the strings A–B. These extra A–B strings give the structure high stiffness in all directions. The diagrams in the figure show the poles first with a small clearance between the poles, to make room for a water tube from the roof to the tent inside (the clearance is also to keep the tension in the A–B strings from being too high). The four poles are anchored to the ground at the four corners. The top of the tent is a square composed of strings. The corners of the square at the top may be positioned relative to the square at the bottom by any angle greater than 45° and less than 90°, where at 90° the poles touch. This angle is dictated by the tension chosen for the strings marked A–B in Figure 1.26. Now install strings (wire, rope or Spectra fiber fishing line) between the nodes of the tent as shown. The strings can be sewn directly into the fabric. The ends of these strings must be placed over the ends of the poles as shown. Such structures can be made deployable by winch power. That is, either by hand or motor one can increase the tension and the length of specified strings to deploy this structure to its final configuration shown in this figure.

1.4.3 Deployable Tensegrity Columns

Insightful papers have been written on the use of symmetry and self-similarity in structure design [Lak93]. Such symmetry offers manufacturing advantages, but the results can also be beautiful (e.g. Eiffel Tower). However, past studies considered fixed structures, whereas we have interest in shape-controllable structures.

Now consider stacking *regular minimal tensegrity prisms* to make a tower or column. Note that a *regular minimal tensegrity prism* is either right-handed or left-handed, depending upon the choice of twist (clockwise or counterclockwise, between the top and bottom polygons). We know of at least two ways of stacking these prisms to form a column. The first method is to stack the prisms alternating between left- and right-handed prisms while allowing the units to overlap. This produces a class 1 tensegrity column of the type that Kenneth Snelson uses in his artwork, see Figure 1.14, and [PMS03, SCS98, SS98b, SCS98, SS03, WdOS06b, WdOS06a, vdWdJ05, Tho00, PRLVC05, PVCL06, KW02b, KW02a, MS02, MS04, IJF06, Fur92]. The second method is to stack only right-handed prisms (or, equivalently, stack only left-handed prisms) without any overlap. This produces a class 2 tensegrity column, which we examine here. Figure 1.27 shows this class 2 column.

Let p denote the number of bars in each prism ($p = 3$ in the figures). Let r and h denote the radius and height of the column, respectively. Let each node at the top be subject to a downward force of equal value, so that the total external force is f. When $f = 0$ we know from our previous discussion that the equilibrium occurs at $\alpha = 30°$. Using the methods of Chapter 2, equilibrium conditions and the minimal mass subject to a buckling constraint

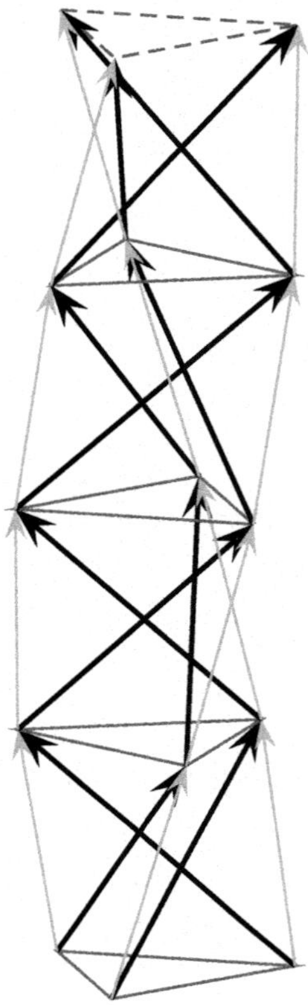

Figure 1.27: A class 2 tensegrity column composed of right-handed *regular minimal tensegrity prisms*

will be calculated in Chapter 3. The total bar mass is minimized by choosing n prisms in the column of height h, where n is given by

$$n^* = \left\lfloor \frac{1}{\sqrt{3(1+\sin(\pi/p)}} \frac{h}{r} \right\rfloor . \tag{1.3}$$

Note that these results do not depend on the mass density, ρ, or Young's modulus, E, of the material, indicating once again that topological optimizations often remain independent of material choice, even though the design of each member (say the diameter of the string or bar) to take the required load will be material dependent.

This example illustrates how minimal mass tensegrity structures can often be characterized by closed form expressions, since the tensegrity definitions allow deep penetration into the mathematical structure of the optimization problems. For a discussion of software and numerical approaches to such problems, see [JKZ98, dOSC06, Cha05, BK88].

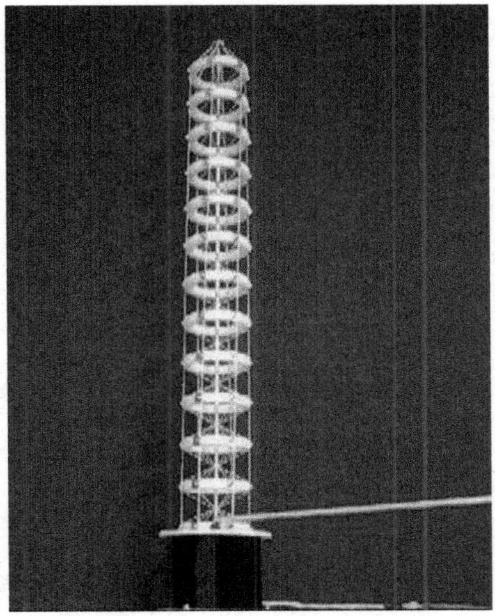

Figure 1.28: A nickel–titanium controlled class 1 tensegrity column, operating with 40 volts and 2 A. The top of the column twists 40°, without changing column height

The column shown in Figure 1.28 uses nickel–titanium wires that reduce their length (about 4%) when a certain current is applied. The vertical strings are stiff and not elastic, whereas the diagonal "strings" are nickel–titanium wires. Sending 2 A of current, at 40 volts, rotates each prism about 4° with respect to the adjacent prism. With 10 stages the twist between the top and the bottom is therefore 40°. The rigid bodies do not touch, hence this is a class 1 tensegrity. The goal of this controlled system is to control the twist from top to bottom, while the height remains constant. One might do this to a tall building to alter sun or wind effects on the heating and air-conditioning of the building. This example also illustrates how a wing might be controlled for roll control, since this structure is exactly the same, topologically, as the structure in Figure 1.32.

1.4.4 Deployable Plates and Antennas

Now consider an assembly of three-bar *regular minimal tensegrity prisms* to make a plate [Han88] or an antenna [MM90, Kni00]. Two different methods of connecting the prisms are illustrated in Figures 1.29 and 1.30. By attaching the ends of the bars of one prism to precisely the correct points along the strings of the adjacent prisms, the stability of the total system is the same

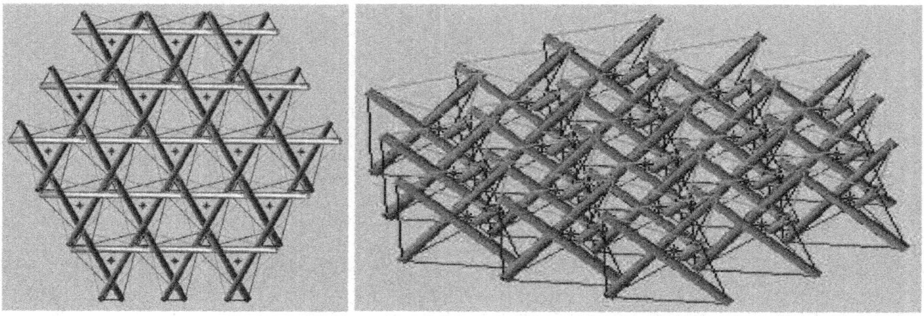

(a) Top view (b) Perspective

Figure 1.29: A plate constructed from *regular minimal tensegrity prisms*

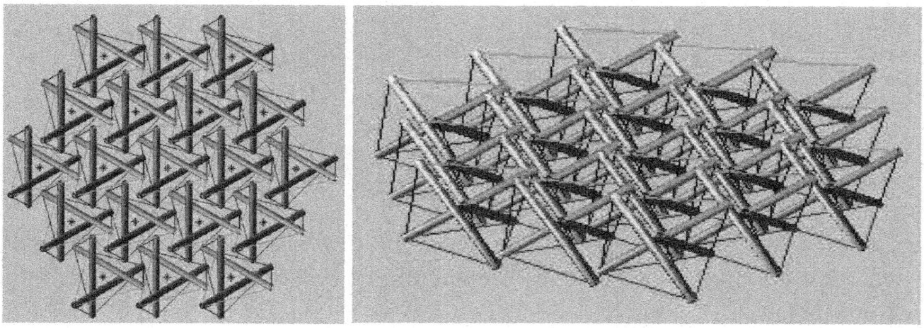

(a) Top view (b) Perspective

Figure 1.30: Another plate constructed from *regular minimal tensegrity prisms*

as for each prism in isolation, namely a twist of 30° between top and bottom triangles.

Since this is a class 1 tensegrity structure, controlling the strings will allow stowing the package into a bundle of parallel bars. Imagine stowing this package into the nose cone of a rocket, and deploying the structure (by string control) in space to form a large flat plate, or a large phased-array antenna. The same stowed package of bars can be controlled to different shapes. Consider that the top and bottom set of three triangular strings each can form an equilateral triangle with a different radius. Figure 1.31 illustrates the case when, for each unit, the top radius is 90% as large as the radius of the bottom [MS04].

In either case, deploying to a flat plate or antenna, a single control function can accomplish the deployment. This plate design is composed of *regular minimal tensegrity prisms* connected in such a way that the unforced equilibrium of the assembly is the same as the equilibrium of each unit. The stable equilibrium of each *tensegrity prism* occurs when the top triangle is rotated

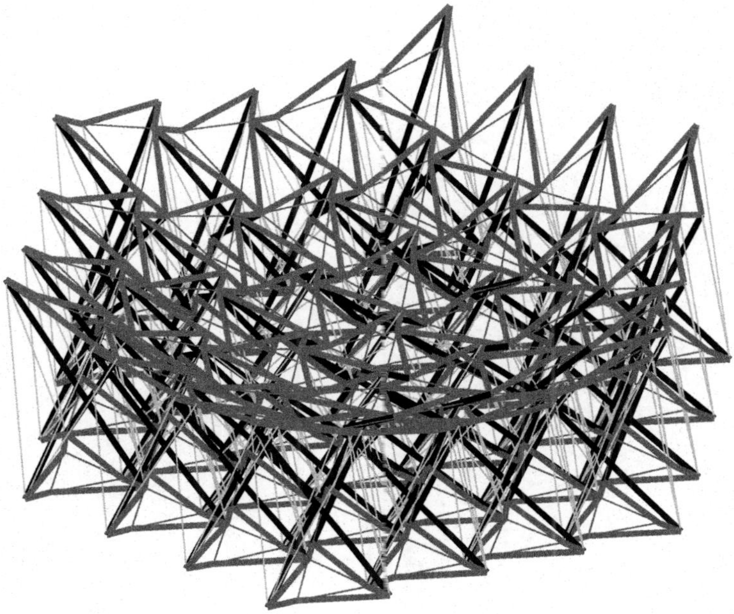

Figure 1.31: The same topology as the flat plate, but radii for top of each prism controlled to 90% of the radii of each prism in the bottom surface

exactly 30° with respect to its bottom triangle and be reminded that this fact remains independent of the radii of either triangle. Hence the radius of both the top and the bottom triangles can be maneuvered arbitrarily and independently, from very small (the stowed configuration) to a larger specified radius (the deployed configuration), while maintaining an equilibrium of the structure at any configuration in between. This is accomplished by simply controlling the three green vertical strings of Figure 1.20, the ones connecting the top and bottom triangles of strings, to maintain 30° twist between top and bottom triangles). This can always be done with a controlled actuation of the vertical strings. In the examples cited here, all units are controlled in exactly the same way, so there is only one control signal to control the deployment of the entire plate.

By adjusting the top radius of each tensegrity prism in this system one can "shape" the antenna to many different configurations that might be useful, say, in controlling the shape of an antenna in space, or controlling the frequency of radiation for which the antenna is "tuned".

The minimal mass plate is considered in Chapter 3, where a given crushing load is applied. Using exactly the same mathematical tools from Chapters 2 and 3 as were used in the minimal mass column in Figure 1.27, we can derive formulas for minimizing the plate composed of a number of *regular minimal tensegrity prisms.*

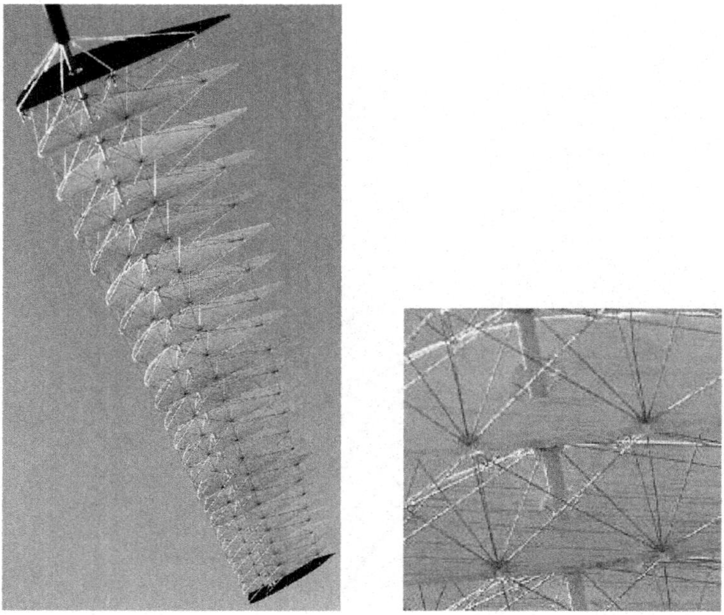

(a) Full Length (b) Close-Up

Figure 1.32: A class 1 tensegrity wing. One rigid body has the shape of the airfoil, and the other type of rigid body has the shape of a long rod (the spar). None of these rigid bodies touch each other

Now finally, consider the task of making a plate made of regular tensegrity prisms stiff. In fact, the topology depicted in Figure 1.31 has extra strings that can make the resulting structure stiff, in the sense there are no soft modes in the structure. It is stiff in all directions. A flat version of Figure 1.31 would render a stiff tensegrity plate, the analysis of which will be detailed in a forthcoming publication. The standard *regular minimal tensegrity prism* in Figure 1.20 is soft in counterclockwise rotation about the axis normal to the page. Mathematically, this means that there is at least one infinitesimal mode (a very small eigenvalue), in addition to the six zero eigenvalues associated with the rigid body motion (see Chapter 2). One needs to add extra strings when high stiffness is required. Recall that our stabilizability conditions only address the stability of the *configuration* of the structure, and not the stability of rigid body motion.

1.4.5 Deployable Wings

Now imagine a one-dimensional deployment of a wing. Figure 1.32 shows photos of a class 1 tensegrity system (this wing model was built by our colleague Carlos Cox). The ribs represent one set of rigid bodies that are not in contact, and the rigid spar (a long rod along the longitudinal axis of

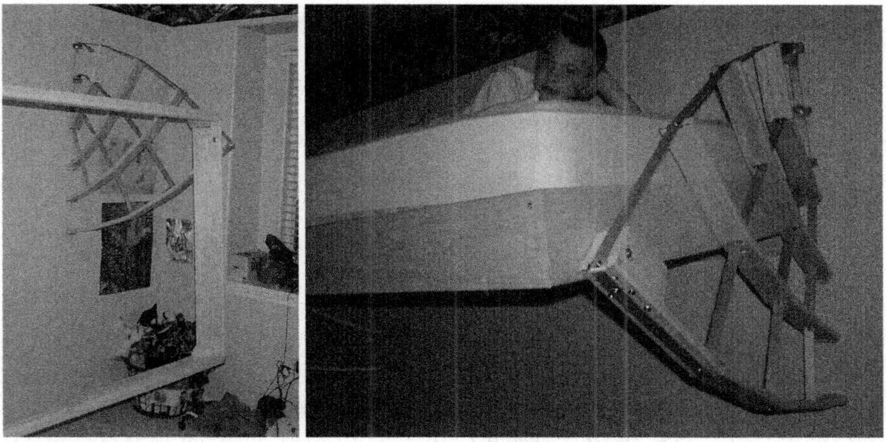

(a) Installation steps (b) Final configuration

Figure 1.33: Wall-mounted bunk bed made from an optimal tensegrity configuration for bending loads

the wing) also makes no contact with the ribs. The wing configuration is stabilized by tensile members shown in the photo. A wing based on these concepts has been designed for deployability from a small stowed package to a fully deployed wing. A preliminary design provided an 80 foot wing that weighs about 0.16 pounds per square foot of wing area and can carry a large weight to very high altitudes. In the actual wing the rigid spar (the rod shown in Figure 1.32) will be replaced by yet another tensegrity structure (optimized minimal mass and compressive strength by the methods in Chapter 3) to deploy the spar to its full length, starting from a stowed package with a very short length. The strings will be controlled to maintain tension during deployment so that the wing can fly at any length in a stable fashion between its stowed length and its fully deployed length. The diagonal strings will be controlled to warp the wing for roll control. Other wing concepts appear in [MTBBS06].

1.4.6 Beds and Broomsticks

Figure 1.33 (provided courtesy of grandson Felix King) shows a bed design that hangs off the wall in a cantilevered manner, so that no connections to the floor are required. The mathematics to minimize mass of a cantilevered structure are derived in Chapter 4, where it is revealed that a class 2 tensegrity structure is the optimal minimal mass solution. In Figure 1.33, the compressive members are small wooden bars connected end to end with flex-pivot joints (a thin wire embedded in the ends of each bar). The tensile members are 1″ nylon straps. With this design the bed can support 1,000 pounds, which is very much over-designed, but adequate for a first prototype supporting a grandson.

The mathematics of Chapter 4 allows one to optimize a structure with *fixed* complexity. That is, the total number of elements (sticks and strings) may be specified a priori. The minimal mass solutions of this chapter approach the infinitely complex (infinite elements leads to a continuum) solution of Michell in [Mic04].

1.4.7 Station-Keeping Buoy

There is much interest in marine tensegrity structures. See the discussions in [WdOS06b, Tri87, Tri91, Luo04, JWL$^+$07a, JWL$^+$07b]. Let us synthesize some requirements for a system of buoys in the deep sea, suitable for weather forecasting or ocean studies. Each buoy is equipped with radio transmitters to send the measured data to ground stations. In the presence of uncertain winds, currents and sea state, it is required that each buoy remains in a fixed position (within an acceptable radius), and operate continuously throughout the life of the components of the system (e.g., 6 months). Each buoy system could to be dropped into the sea through a small hole in the belly of an airplane. The buoy system must be stowed into a small package to pass through the hole in the airplane, and once the package hits the ocean, it must deploy to its fully operational configuration. Obviously the system must have propulsion and maneuvering capability to perform the station-keeping function. To operate the electronic equipment and to power a propulsion system, such a system must obviously generate power from the local environment (wind, sun, or water), since no external source of power is available.

Current buoy systems can do all of these tasks except the energy conversion, the station keeping, and the long life. Currently they are battery powered, are uncontrolled, and have a very short life. Hence, the wind and sea carry the current buoy away from the intended location.

A tensegrity solution to the structural part of this problem has been created, built, and tested in sea trials. The system is designed to convert wave motion to electrical energy. The instrumented buoy system is shown in Figure 1.34 (note the generator attached to the bottom of the buoy).

The structural system configuration is a five-bar, eight-string, class 2 tensegrity system sketched in Figures 1.35 and 1.36. The four strings (certain red thin lines) that lie in the horizontal plane are actually edges of a membrane in tension, shown in Figures 1.36 and 1.37). We will label this system the *Wave-Powered Station-Keeping Buoy* (WPSB). The vertical red cable in Figures 1.35 and 1.36 wraps around the generator rotor and then drops to the WPSB below.

The buoy follows the surface of the sea wave, but the WPSB resists vertical motion because of the mass and the membrane. This wave induced elongation of the vertical cable rotates the generator, creating power for immediate use to drive the motorized vehicle, power the electronics, operate the

Figure 1.34: The instrumented buoy for the wave-powered station-keeping Buoy (WPSB)

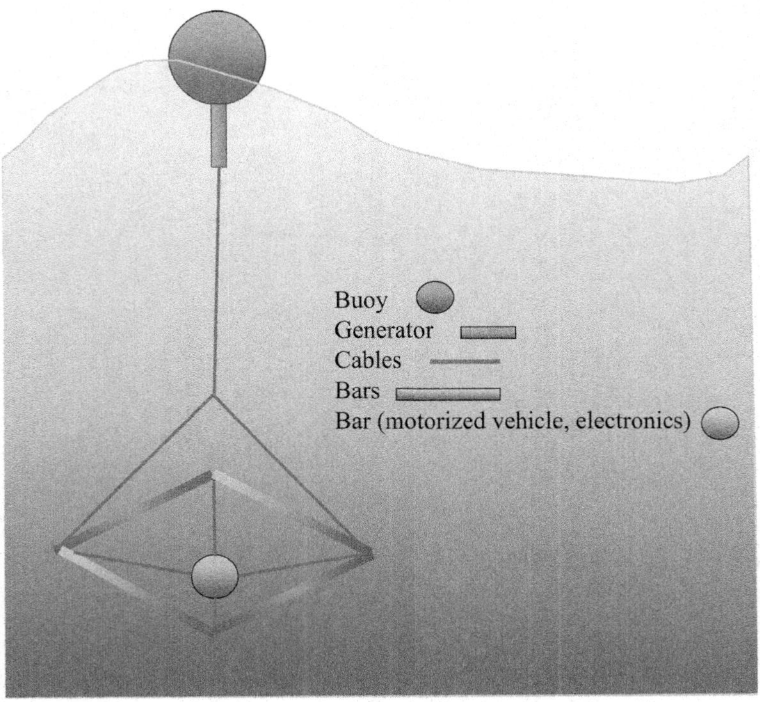

Figure 1.35: The first wave-powered vehicle. A class 2 tensegrity system for a wave-powered station-keeping buoy (WPSB)

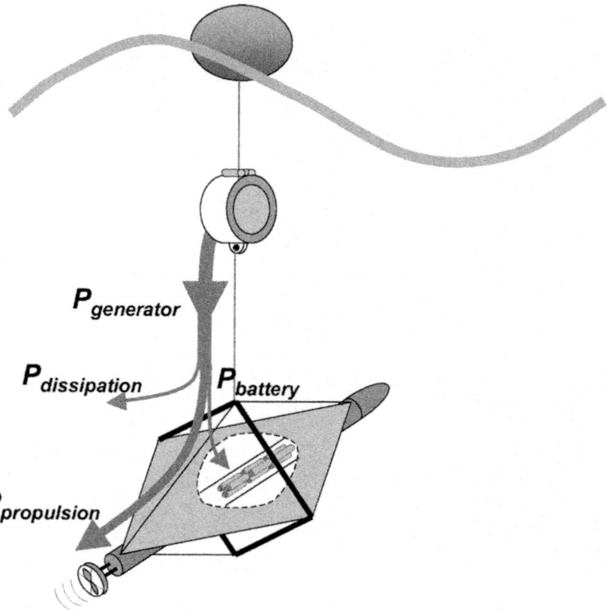

Figure 1.36: The functional diagram of the wave-powered station-keeping buoy

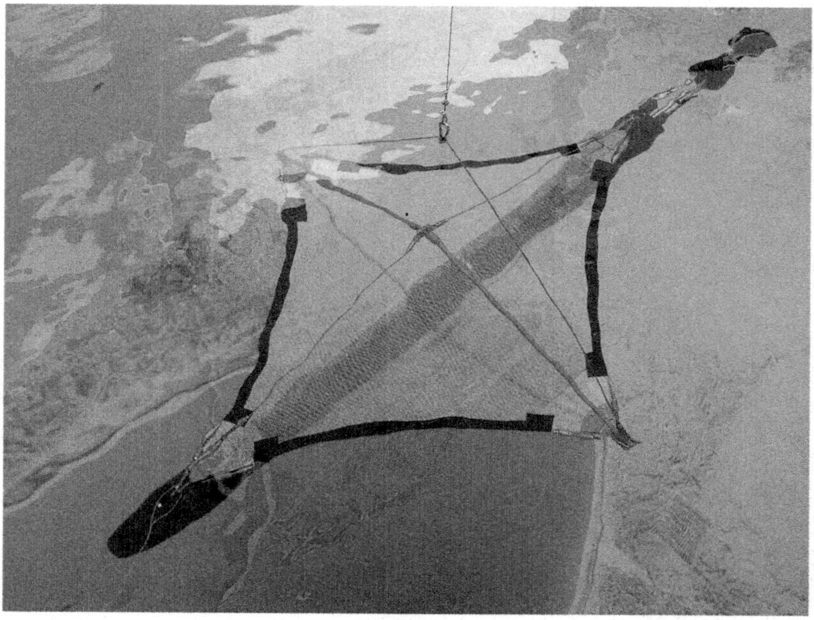

Figure 1.37: A test of the WPSB in a pool

navigation system, etc. This power is routed interior to the stainless steel vertical cable, travels down the cable to the WPSB, and powers the electric operations (propulsion and guidance systems) inside the WPSB.

Figure 1.37 shows the non-deployable prototype, where all bars were aluminum and the cables were stainless steel. The propulsion system is a thruster mounted on the rear of the WPSB. The electronics, sensors, magnetic compass, and guidance and control computers are inside the horizontal aluminum tube of Figure 1.37. This aluminum tube serves as one of the bars in the tensegrity design. The goal was to keep the buoy within 250 meters of the commanded coordinate, in sea state 2. Due to severe weather, the system was tested in sea state 5. Yet, in the presence of high sea state, high winds, and currents of 1 m/s the control system kept the buoy within 60 m of the target for 2 hours. The tensegrity concept for the project proved feasible, and this project provided the first WPSB, although a very slow submarine indeed, traveling at 2 m/s. To demonstrate feasibility of the mechanical parts of the system, this first test used a prototype that did not deploy, used a dummy generator (mechanical motions correct) and was battery-powered. The deployment and the power generator are yet to be designed. Such objectives are discussed further in [SS06].

1.4.8 Dynamics of Tensegrity Systems

Perhaps the main reason tensegrity systems have remained the tool of the artist rather than the tool of the engineer is the absence of tools to efficiently write and solve the equations of motion, i.e., the *dynamics* of tensegrity systems. As seen before in this chapter, tensegrity systems often have a large number of components and one needs to carefully keep track of all the variables involved. This difficulty will be overcome in Chapter 5, where a systematic and efficient formulation of the equations of motion of tensegrity systems will be presented.

The equations of motion are provided in simple form, which make computation and control design (to be addressed later in Chapter 6) easier. The formulation encompasses constrained and unconstrained tensegrity systems, providing non-minimal yet simple realizations of the constrained dynamics, and finally a new matrix form of the equations of motion for class 1 tensegrity systems. By using the vector along the rod as a generalized coordinate, the final equations are devoid of the transcendental functions that complicate the form of the dynamics.

1.4.9 Control of Tensegrity Systems

There does not exist a *systems design theory*, where all components of a system are designed to efficiently work together to accomplish a goal. Control theory as we know it today allows solutions only *after* the critical decisions are made about component designs. This book attempts small steps to develop scientific tools to relate the overall motion requirements to the initial design of the structure and control.

The main motivation for the use of tensegrity structures is the advantages of integrating control functions within the design of the structure. Traditionally, a structure would be designed by a finite element method or a gridding method [Ben89, BK88, DB92, Ben95, JKZ98, BS03]. In both of these cases the structure has a fixed equilibrium, since the joints are welded or pinned. At present, the control function is added to the structural system by mounting (on the designed structure) control actuators, to make momentary or permanent changes in the shape, pressing against the equilibrium of the structure. This forced change in shape requires energy and perhaps large forces. In other words the actuators "torture" the structure to do things the structure does not naturally want to do. The idea of controlled tensegrity systems is to allow the dynamics of the controller and the dynamics of the structure to *cooperate* more fully. This is accomplished by using control to *change* the equilibrium of the structure, rather than pressing against a fixed equilibrium of the structure.

Figure 1.20 provides the simplest opportunity to describe control problems, which will be formally addressed in Chapter 6. Suppose sensors measure the rate (v) at which the green vertical string changes length. Suppose the tension (u) in this string is controlled by a feedback control algorithm. If, for example, one wishes to change the diameter or height, or stiffness of the regular tensegrity prism, then all the green vertical strings could be controlled in exactly the same manner, increasing length or decreasing the tension or length of the green vertical strings until the desired height is reached, or until the right stiffness is obtained.

Other tensegrity control examples are discussed in [DMPC98, dJS01, dJSM01, CABS04, MS06, MS05, WdOS06b, WdOS06a, WdOS09].

Figure 1.38 illustrates a column (drawn on its side) exposed to a compressive load, which we have equivalently transferred to a prestress, $t = 50$. Now suppose we wish to calculate the optimal number of units p (each having two bars and four strings) to put in the column of a given height L and width D. Subject to buckling constraints the optimal number is the closest integer to the value $p = L/D\sqrt{2/3}$, where the length of the column is L and the width is D.

Now suppose this structure should be controlled so that the variance of the vertical displacement y at the top right corner should be less than 0.01. That is, $E(y^2) < 0.01$, where the disturbance is a zero-mean white noise acting on the upper right-hand corner of the structure in Figure 1.39. The intensity of the noise is 10. All the vertical strings will be controlled by state feedback in this example. Hence the number of control actuators (controlling vertical string tensions) increase as p increases (p is the number of self-similar units used in the final design of the column). Superimposed on the control signal from each actuator is actuator noise, which in this example, is zero-

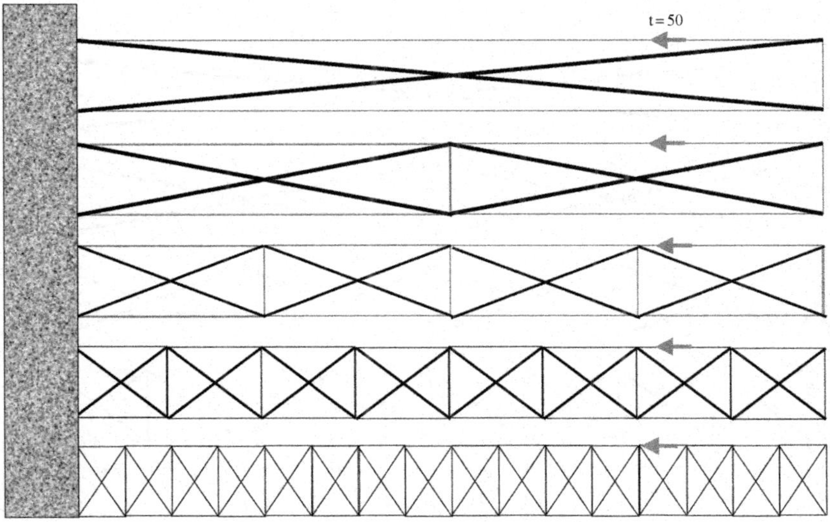

Figure 1.38: A column with an arbitrary number of two-bar units. The minimal mass is obtained with p units, where $p = L/D\sqrt{2/3}$

mean white noise with intensity 10^{-6}. The total control energy required to accomplish the objective $E(y^2) < 0.01$ is minimized for *each* of the column designs shown in Figure 1.39 for $p = 1, 2, 3 \ldots$. Then the control energy required for each of these designs is compared, yielding $p = 4$ as the structure that requires the least control energy for the given performance requirement $E(y^2) < 0.01$. But $p = 4$ is not the minimal mass structure, illustrating that the structure and control design problems are not coordinated, in any way. Now suppose we introduce another free parameter in the control design problem, by letting the actuator noise be free to choose, in addition to the control signals. Assuming that actuator cost is inversely proportional to the noise intensity of the actuator, the mathematical freedom to choose the actuator noise is equivalent choosing the cost of the actuator. The following problem can then be solved: *Find the minimal cost of all actuators, over the set of all self-similar structures (columns), so that the minimal energy required to deliver the closed-loop performance ($E(y^2) < 0.01$) corresponds to the structure that has minimal mass.* It is not difficult to show that one can always choose such an actuator noise, intensity and in this example that value is 2×10^{-7}. Notice that it would be a waste of money to make the actuator better (less noisy and more costly) than this number, since then the minimal energy for controlling the column would occur for a structure that was more complex and more massive than our "optimal" design above. Note that this is not a globally optimal structure design, but only one that uses self-similar structures.

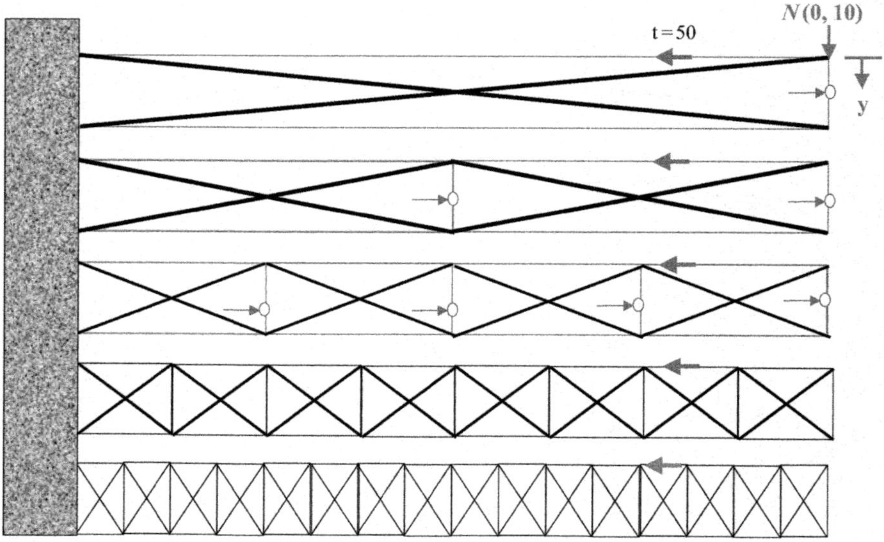

Figure 1.39: A column with a specified number (p) of two-bar units under feedback control. The control energy is minimized while keeping the vibrations less than $E(y^2) < 0.01$. The structure requiring the smallest energy among all the structures is for $p = 4$ and $p = 8$ for actuator noises, respectively, 10^{-6} and 2×10^{-7}

This is an example where the design of the structure, the design of the control system, the design (hence cost) of the instruments (actuators and sensors), and the choice of the number of actuators are all combined.

The photo in Figure 1.40 and the sketch in Figure 1.41 shows the experimental setup for a vibration control problem, using three actuators, three sensors, and decentralized control for a tensegrity column composed of nine bars. The objective is to reduce the vibration at the top of the structure, while actuators and sensors are at the bottom of the structure. Decentralized control in this example allows each colocated sensor/actuator pair to communicate in a closed-loop control loop, without any coordination with the other sensors, actuators, or controllers. This type of control would be advantageous for a large complex system, such as a deployable tensegrity system. Figure 1.42 shows the frequency response of this system being mounted on a shaker table to simulate a force spectrum similar to an earthquake. For different values of control gains, this system demonstrates the ability to suppress the vibration by almost 30 dB. The main point of these plots is to demonstrate a certain accuracy of the tensegrity model. Usually in vibration control problems, the inaccuracies of the model upon which the control system is based manifest themselves by driving the closed-loop system unstable for high gains, because the frequencies of the modes of the model

Figure 1.40: Feedback control of a nine-bar tensegrity column in the presence of a shaker table. Three actuators and three sensors located at the bottom of the structure are used to control the vibrations at the top of the structure

are usually not the same as the frequencies of the modes of the hardware. Note from Figure 1.42 that the open loop system and the closed-loop system match the frequencies of the real system for all of the dominant modes of the structure.

The thesis of Jack Aldrich [Ald04] showed how one might control two read/write heads at once by designing a disk drive suspension system with two heads and mounting them in a fashion to create a class 1 tensegrity structure. Figure 1.43 shows two desired trajectories, one for each read/write head. The goal is to reach the end of this specified path in minimal time, subject to the constraint that no string becomes slack at any instant of time (hence a lower tension bound of 5 is specified), and no string yields (hence an upper bound for the instantaneous string force is set at 95), no bar buckles (so an upper bound of the bar force is added as well). The solution involves bang–bang control with two switches from minimum to maximum string force during the optimal control along the given path. The green lighter sections

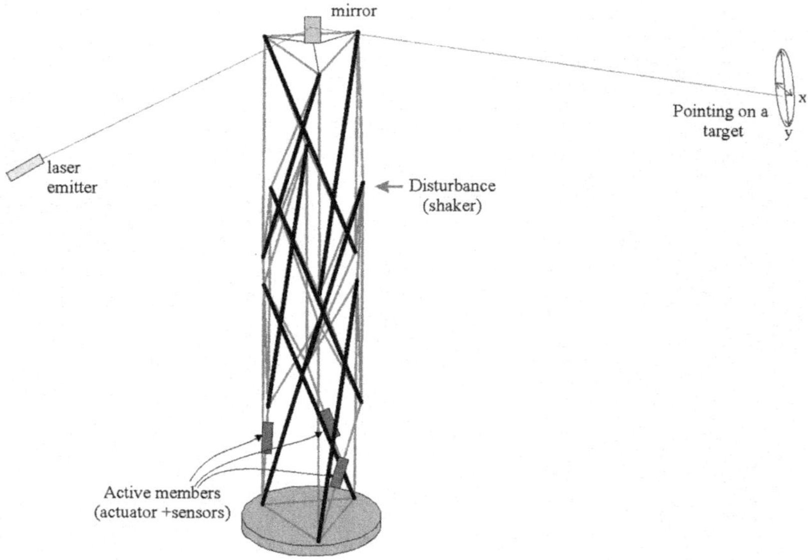

Figure 1.41: The Experimental setup for vibration control of a tensegrity column

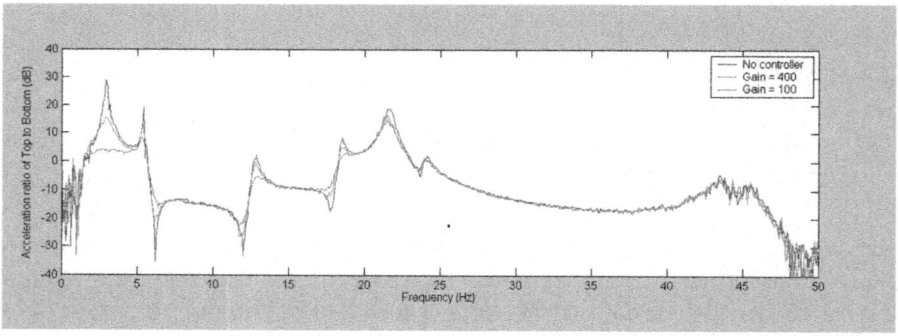

Figure 1.42: The frequency response of the closed-loop system with three values of control gain, including 0 for open loop. The red curve reduces the vibration of the first mode by almost 30db

of the paths represent where at least one string is at its maximum allowable force (95), while the red darker section of the path corresponds to at least one string being at its lower allowable force (5). The figure only shows the string forces at a frozen moment in time.

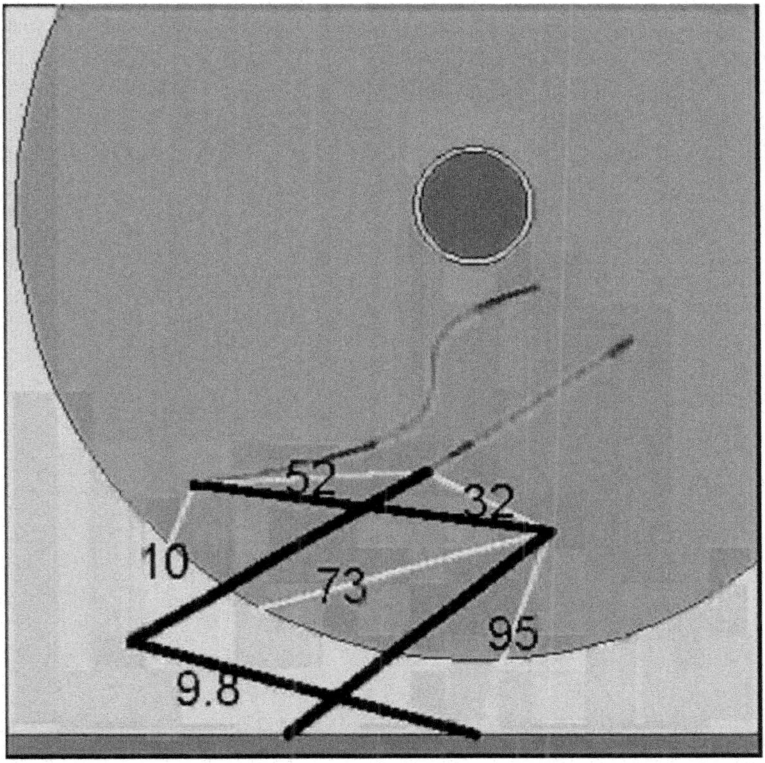

Figure 1.43: A disk drive with two heads, traveling along specified paths to reach the final position in minimal time. The numbers represent the string tensions at a frozen moment in time. On the *red* light part of the path at least one string is at its smallest allowable tension. On the *green* darker path at least one string is at its maximal allowable tension

1.5 Chapter Summary

The purpose of this chapter is to introduce tensegrity without much mathematics, to give some insightful examples to show how biology (red blood cells, spider fiber, animal skeletal systems for locomotion) and art (Ioganson, 1921, Kenneth Snelson 1948, and Buckminister Fuller) have exploited the beneficial and sometimes beautiful topologies of material for different purposes. Motivation is provided suggesting the need for the development of more advanced mathematical machinery to allow scientists and engineers to analyze and design tensegrity systems that can achieve shape control and other engineering functions beyond current capabilities. This chapter provides a guide to the later chapters of the book, illustrating certain minimal mass results that will be derived in later chapters (minimal mass solutions for symmetric columns and plates). This chapter provides an easily digested treatment of tensegrity without math, but it is not exhaustive in that regard. It does not point to many results that can only be motivated within the context of the detailed chapters. One such topic is optimal complexity, a theme prominent in Chapters 3 and 4, where great emphasis is placed on making a structure not more complicated than necessary to minimize mass.

Chapter 2

Analysis of Static Tensegrity Structures

This chapter develops convenient mathematical algorithms to characterize stable equilibria of tensegrity systems, to characterize stiffness properties, to characterize the relationships between force and configuration, and to characterize the mass of the system, subject to buckling or yield constraints. The final design of a tensegrity system will involve both the static and dynamic properties. Often the dynamic motion can be greatly improved by changing design parameters (the types of building blocks, the configuration of each unit, and the configuration of the whole network of elements) that were fixed in the static analysis of this chapter. Hence, we are interested in efficient computational tools that will allow rapid and efficient changes, whose impact will later, in Chapter 5, be quantified from dynamic properties.

2.1 Nodes, Members, and Connectivity

Let n be the number of nodes in a structure given by the three-dimensional vectors

$$\mathbf{n}_i \in \mathbb{R}^3, \quad i = 1, \ldots, n.$$

The configuration of the entire structure is described by the node vector

$$\mathbf{n} = \begin{pmatrix} \mathbf{n}_1 \\ \vdots \\ \mathbf{n}_n \end{pmatrix} \in \mathbb{R}^{3n}. \tag{2.1}$$

Let $\mathbf{m}_k$ be the vector that describes the kth member in the structure whose nodes are $\mathbf{n}_{i_k}$ and $\mathbf{n}_{j_k}$, i.e.,

$$\mathbf{m}_k := \mathbf{n}_{i_k} - \mathbf{n}_{j_k}.$$

R.E. Skelton, M.C. de Oliveira, *Tensegrity Systems*,
DOI 10.1007/978-0-387-74242-7_2, © Springer Science+Business Media, LLC 2009

If m is the total number of members in the structure, the set of members of the structure is described by the element vector

$$\mathbf{m} = \begin{pmatrix} \mathbf{m}_1 \\ \vdots \\ \mathbf{m}_n \end{pmatrix} \in \mathbb{R}^{3m}.$$

Now define the vector $\mathbf{e}_i \in \mathbf{R}^n$ as the vector with one in the ith position and zero elsewhere. Let

$$\mathbf{d}_k := \mathbf{e}_{i_k} - \mathbf{e}_{j_k}.$$

Then define the matrix

$$\mathbf{N} = \begin{bmatrix} \mathbf{n}_1 & \cdots & \mathbf{n}_n \end{bmatrix} \in \mathbb{R}^{3\times n}. \tag{2.2}$$

It follows that

$$\mathbf{m}_k = \mathbf{N}\left(\mathbf{e}_{i_k} - \mathbf{e}_{j_k}\right) = \mathbf{N}\,\mathbf{d}_k.$$

Define the matrix

$$\mathbf{D} = \begin{bmatrix} \mathbf{d}_1 & \cdots & \mathbf{d}_m \end{bmatrix} \in \mathbb{R}^{n\times m}.$$

The matrix obtained by transposition of $\mathbf{D}$, i.e.,

$$\mathbf{C} = \mathbf{D}^T \in \mathbb{R}^{m\times n},$$

is called the *connectivity matrix*. This matrix is made of ones, zeros, and minus ones and since the rows of $\mathbf{C}$ are vectors of the form $\mathbf{d}_k$ the sum of all rows of $\mathbf{C}$ must be zero, i.e.,

$$\mathbf{C}\mathbf{1}_n = \mathbf{0},$$

where $\mathbf{1}_n \in \mathbb{R}^n$ is a vector where all entries are ones.

If we define a matrix of member vectors

$$\mathbf{M} = \begin{bmatrix} \mathbf{m}_1 & \cdots & \mathbf{m}_m \end{bmatrix} \in \mathbb{R}^{3\times m},$$

then

$$\mathbf{M} = \mathbf{N}\mathbf{C}^T. \tag{2.3}$$

Using the standard vec operator we have the following definition.

Definition 2.1 *For any matrix* $\mathbf{X} \in \mathbb{R}^{m\times n}$ *the* vec *operation produces the column vector* $\mathbf{x} \in \mathbb{R}^{mn}$, $\mathbf{x} = \text{vec}(\mathbf{X})$ *by stacking up the columns of* $\mathbf{X}$.

Hence, we have the facts:

$$\mathbf{n} = \text{vec}(\mathbf{N}), \qquad \mathbf{m} = \text{vec}(\mathbf{M}). \tag{2.4}$$

It also follows from (2.3) that

$$\mathbf{m} = \text{vec}(\mathbf{m}) = (\mathbf{C} \otimes \mathbf{I}_3)\ \text{vec}(\mathbf{n}),$$

where $\otimes$ denotes the Kronecker product, defined below.

Definition 2.2 *Let* $\mathbf{A}$ *and* $\mathbf{B}$ *be* $a \times b$ *and* $k \times n$*, respectively. The notation* $\mathbf{A} \otimes \mathbf{B}$ *defines a* $ak \times bn$ *matrix such that the* ij *block of matrix* $\mathbf{A} \otimes \mathbf{B}$ *is a* $k \times n$ *matrix defined by*

$$(\mathbf{A} \otimes \mathbf{B})_{ijblock} = A_{ij}\mathbf{B}. \tag{2.5}$$

See [HJ85] for additional properties of Kronecker products.

2.2 Potential and Force

Suppose now that the kth structural member is associated with the potential energy

$$V_k(\mathbf{n}) = V_k(\|\mathbf{m}_k\|),$$

which is a function exclusively of the length of $\mathbf{m}_k$. The total potential energy on the structure is

$$V(\mathbf{n}) = \sum_{k=1}^{m} V_k(\|\mathbf{m}_k\|).$$

The total force at the nodes due to the potential $V(\mathbf{n})$ is the vector

$$\mathbf{f}(\mathbf{n}) = -\partial_{\mathbf{n}} V(\mathbf{n}) \in \mathbb{R}^{3n}.$$

This vector can be computed by noting that

$$\partial_{\mathbf{n}_{i_k}} \|\mathbf{m}_k\| = \partial_{\mathbf{n}_i} \|\mathbf{n}_{i_k} - \mathbf{n}_{j_k}\| = \frac{\mathbf{n}_{i_k} - \mathbf{n}_{i_k}}{\|\mathbf{n}_{i_k} - \mathbf{n}_{j_k}\|} = \frac{\mathbf{m}_k}{\|\mathbf{m}_k\|},$$

so that

$$\partial_{\mathbf{n}_{ik}} V_k(\|\mathbf{m}_k\|) = \sigma_k(\|\mathbf{m}_k\|)\, \mathbf{m}_k,$$

where the scalars

$$\sigma_k(\|\mathbf{m}_k\|) := \frac{V_k'(\|\mathbf{m}_k\|)}{\|\mathbf{m}_k\|}$$

are known as *force densities*, and

$$V_k'(\|\mathbf{m}_k\|) = \frac{d\,V_k(\|\mathbf{m}_k\|)}{d\,\|\mathbf{m}_k\|}.$$

Since

$$\partial_{\mathbf{n}_{j_k}} V_k(\|\mathbf{m}_k\|) = -\partial_{\mathbf{n}_{i_k}} V_k(\|\mathbf{m}_k\|) = -\sigma_k(\|\mathbf{m}_k\|)\,\mathbf{m}_k,$$

we have

$$\mathbf{f}_k(\mathbf{n}) = -\partial_{\mathbf{n}} V_k(\|\mathbf{m}_k\|) = -\sigma_k(\|\mathbf{m}_k\|)\,(\mathbf{d}_k \otimes \mathbf{m}_k)\,,$$

where $\mathbf{d}_k = \mathbf{e}_{i_k} - \mathbf{e}_{j_k}$. From the above formulas we conclude that

$$\mathbf{f}(\mathbf{n}) = \sum_{k=1}^{m} \mathbf{f}_k(\mathbf{n}).$$

Note that because

$$\mathbf{d}_k \otimes \mathbf{m}_k = \mathrm{vec}(\mathbf{m}_k \mathbf{d}_k^T),$$

it is convenient to define

$$\mathbf{F}_k(\mathbf{n}) = -\sigma_k(\|\mathbf{m}_k\|)\,\mathbf{m}_k \mathbf{d}_k^T \in \mathbb{R}^{3\times n},$$

and the force matrix

$$\mathbf{F}(\mathbf{n}) = \sum_{k=1}^{m} \mathbf{F}_k(\mathbf{n}).$$

Defining the diagonal matrix of force densities

$$\mathbf{\Sigma}(\mathbf{m}) = \mathrm{diag}\left[\sigma_1(\|\mathbf{m}_1\|), \ldots, \sigma_m(\|\mathbf{m}_m\|)\right] \in \mathbb{R}^{m\times m},$$

we have

$$\begin{aligned}
\mathbf{F}(\mathbf{n}) &= \begin{bmatrix}\mathbf{m}_1 & \cdots & \mathbf{m}_m\end{bmatrix} \mathbf{\Sigma}(\mathbf{m}) \begin{bmatrix}\mathbf{d}_1^T \\ \vdots \\ \mathbf{d}_m^T\end{bmatrix}, \\
&= \mathbf{M}\mathbf{\Sigma}(\mathbf{m})\mathbf{C}, \\
&= \mathbf{N}\mathbf{C}^T\mathbf{\Sigma}(\mathbf{m})\mathbf{C}.
\end{aligned}$$

Of course

$$\mathbf{f}(\mathbf{n}) = \mathrm{vec}(\mathbf{F}(\mathbf{n})).$$

2.3 Linear Springs and Strings

For most part of this book we will model bars as rigid members or linear springs and tendons as linear strings.

The potential energy associated with linear springs is of the form

$$\begin{aligned} V(\|\mathbf{m}\|) &= \frac{1}{2}\kappa\left(\|\mathbf{m}\| - l^0\right)^2, \\ V'(\|\mathbf{m}\|) &= \kappa\left(\|\mathbf{m}\| - l^0\right), \\ V''(\|\mathbf{m}\|) &= \kappa, \end{aligned}$$

where $\mathbf{m}$ is the vector describing the spring and the scalar $l^0 > 0$ is known as the *rest length* of the spring.

Note also that

$$\sigma(\|\mathbf{m}\|) = \kappa\left(1 - \frac{l^0}{\|\mathbf{m}\|}\right). \tag{2.6}$$

Indeed, V, V', and V'' can all be parametrized as a function of σ and κ since

$$\begin{aligned} V(\|\mathbf{m}\|) &= \frac{1}{2\kappa}\sigma^2\|\mathbf{m}\|^2, \\ V'(\|\mathbf{m}\|) &= \sigma\|\mathbf{m}\|, \\ V''(\|\mathbf{m}\|) &= \kappa, \end{aligned}$$

a fact that we will explore in the next chapters. Note that in this case l^0 can be recovered by inverting (2.6) as

$$l^0 = \|\mathbf{m}\|\left(1 - \frac{\sigma}{\kappa}\right).$$

Note that for $l^0 > 0$ we must have

$$\kappa > \sigma. \tag{2.7}$$

For strings, the expressions are the same except that $V(\mathbf{n})$ is null when $\|\mathbf{m}\| < l^0$. That is,

$$V(\|\mathbf{m}\|) = \begin{cases} \frac{1}{2}\kappa\left(\|\mathbf{m}\| - l^0\right)^2, & \|\mathbf{m}\| < l^0 \\ 0, & \|\mathbf{m}\| \geq l^0 \end{cases}.$$

Note that $V(\|\mathbf{m}\|)$ is not differentiable at $\|\mathbf{m}\| = l^0$. If V' and V'' are needed at $\|\mathbf{m}\| = l^0$ we use the "right derivative", i.e., $V'(\|\mathbf{m}\|) = 0$ and $V''(\|\mathbf{m}\|) = \kappa$.

2.4 Equilibrium

The force vector is associated with a Taylor series expansion of $V(\mathbf{n})$ in the direction $\mathbf{h}$ in the form

$$V(\mathbf{n} + \epsilon\mathbf{h}) = V(\mathbf{n}) - \epsilon\mathbf{h}^T\mathbf{f}(\mathbf{n}) + \frac{\epsilon^2}{2}\mathbf{h}^T\mathbf{K}(\mathbf{n})\mathbf{h} + O(\epsilon^3). \tag{2.8}$$

The second-order coefficient of the above series, the symmetric matrix $\mathbf{K}(\mathbf{n})$, is known as the *stiffness matrix* and will be studied in the next section.

A mechanical system with configuration $\mathbf{n}$ and potential energy function $V(\mathbf{n})$ is said to be in equilibrium at $\bar{\mathbf{n}}$ if

$$\mathbf{f}(\bar{\mathbf{n}}) = -\partial_{\mathbf{n}} V(\bar{\mathbf{n}}) = \mathbf{0},$$

that is, if $\bar{\mathbf{n}}$ is a stationary point of the potential function.

In tensegrity structures it is also important to qualify the nature of the force on each member. For that sake, let us partition

$$\mathbf{M} = \begin{bmatrix}\mathbf{B} & \mathbf{S}\end{bmatrix} \in \mathbb{R}^{3\times(m_b+m_s)},$$

where $\mathbf{B}$ describes bar elements and $\mathbf{S}$ describes string elements only. Partition the diagonal matrix $\mathbf{\Sigma}$ in the form

$$\mathbf{\Sigma} = \begin{bmatrix}-\mathbf{\Lambda} & \mathbf{0}\\ \mathbf{0} & \mathbf{\Gamma}\end{bmatrix} \in \mathbb{R}^{3\times(m_b+m_s)},$$

where we flipped the sign of $\mathbf{\Lambda}$.

A string can only carry tension, which means

$$\mathbf{\Gamma} \succeq 0.$$

Recall that $\mathbf{\Gamma}$ (and $\mathbf{\Lambda}$) are diagonal matrices. Therefore, the matrix inequality $\mathbf{\Gamma} \succeq 0$ is equivalent to the vector entrywise inequality $\gamma \geq 0$.

A bar will usually carry only compression, hence $\mathbf{\Lambda} \succeq 0$. However, we shall not impose that $\mathbf{\Lambda} \succeq 0$, because a bar can indeed carry tension, although not as efficiently as a string. Therefore, if $\lambda_i < 0$ for the ith bar it can be advantageous to replace it by a string member.

Using the matrix form of the force $\mathbf{F}$, a complete statement of equilibrium for a tensegrity structure is then

$$\mathbf{F} = \begin{bmatrix}-\mathbf{B}\mathbf{\Lambda} & \mathbf{S}\mathbf{\Gamma}\end{bmatrix}\mathbf{C} = \mathbf{0}, \qquad \mathbf{\Gamma} \succeq 0.$$

We often write the inequality constraint in terms of the vector γ in the equivalent form:

$$\begin{bmatrix}-\mathbf{B}\mathbf{\Lambda} & \mathbf{S}\mathbf{\Gamma}\end{bmatrix}\mathbf{C} = \mathbf{0}, \qquad \gamma \geq 0.$$

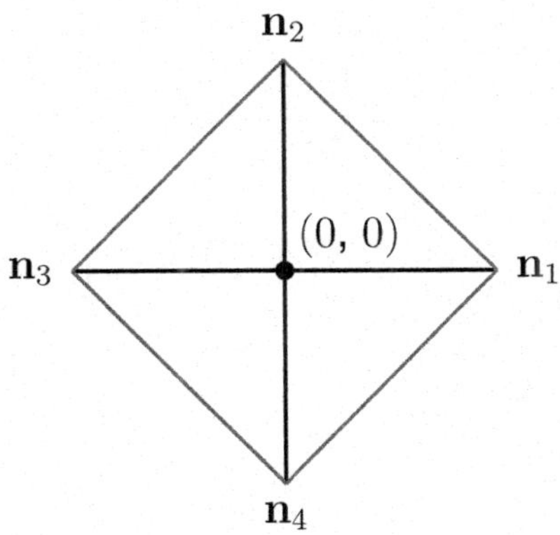

Figure 2.1: Planar tensegrity cross

Because $\mathbf{C}\mathbf{1}_n = \mathbf{0}$, any one column of $\mathbf{C}$ can be dropped without loss of generality, and this would reduce by three the total number of equations to be solved. This device can and should be used to improve the behavior of numerical procedures to compute equilibrium solutions.

When dealing with tensegrity structures it is convenient to partition the connectivity matrix $\mathbf{C}$ as

$$\mathbf{C} = \begin{bmatrix} \mathbf{C}_B \\ \mathbf{C}_S \end{bmatrix},$$

such that

$$\begin{bmatrix} \mathbf{B} & \mathbf{S} \end{bmatrix} = \mathbf{N} \begin{bmatrix} \mathbf{C}_B^T & \mathbf{C}_S^T \end{bmatrix}.$$

That is, $\mathbf{C}_B$ describes the "bar connectivity" and $\mathbf{C}_S$ describes that "string connectivity".

Example 2.1

Consider the planar tensegrity cross structure with four ($n = 4$) nodes

$$\mathbf{n}_i = \mathbf{R}(\pi/2)^{i-1}\,\mathbf{e}_x, \qquad i = 1, \ldots, 4,$$

as shown in Figure 2.1. The structure has six ($m = 6$) members, two bars (ir red) and four strings (in black)

$$\begin{aligned} \mathbf{b}_i &= \mathbf{n}_{i+2} - \mathbf{n}_i, & i &= 1, 2, \\ \mathbf{s}_i &= \mathbf{n}_{[i \pmod 4]+1} - \mathbf{n}_i, & i &= 1, \ldots, 4. \end{aligned}$$

Let

$$\begin{aligned} \mathbf{N} &= \begin{bmatrix} \mathbf{n}_1 & \cdots & \mathbf{n}_4 \end{bmatrix} \in \mathbb{R}^{2\times 4}, \\ \mathbf{B} &= \begin{bmatrix} \mathbf{b}_1 & \mathbf{b}_2 \end{bmatrix} \in \mathbb{R}^{2\times 2}, & \mathbf{S} &= \begin{bmatrix} \mathbf{s}_1 & \cdots & \mathbf{s}_4 \end{bmatrix} \in \mathbb{R}^{2\times 4}. \end{aligned}$$

The connectivity matrix is

$$\mathbf{C} = \begin{bmatrix} \mathbf{C}_B \\ \mathbf{C}_S \end{bmatrix} \in \mathbb{R}^{6\times 4},$$

where

$$\mathbf{C}_B = \begin{bmatrix} -1 & 0 & 1 & 0 \\ 0 & -1 & 0 & 1 \end{bmatrix}, \qquad \mathbf{C}_S = \begin{bmatrix} -1 & 1 & 0 & 0 \\ 0 & -1 & 1 & 0 \\ 0 & 0 & -1 & 1 \\ 1 & 0 & 0 & -1 \end{bmatrix}.$$

Note that

$$\mathbf{N} = \begin{bmatrix} \mathbf{e}_x & \mathbf{e}_y & -\mathbf{e}_x & -\mathbf{e}_y \end{bmatrix} = \begin{bmatrix} \mathbf{I}_2 & -\mathbf{I}_2 \end{bmatrix}$$

and

$$\begin{aligned} \mathbf{M} &= \begin{bmatrix} \mathbf{B} & \mathbf{S} \end{bmatrix}, \\ &= \mathbf{N}\mathbf{C}^T, \\ &= \begin{bmatrix} \mathbf{I}_2 & -\mathbf{I}_2 \end{bmatrix} \begin{bmatrix} -1 & 0 & -1 & 0 & 0 & 1 \\ 0 & -1 & 1 & -1 & 0 & 0 \\ 1 & 0 & 0 & 1 & -1 & 0 \\ 0 & 1 & 0 & 0 & 1 & -1 \end{bmatrix}, \\ &= \begin{bmatrix} -2 & 0 & -1 & -1 & 1 & 1 \\ 0 & -2 & 1 & -1 & -1 & 1 \end{bmatrix}. \end{aligned}$$

The structure will be in equilibrium if

$$\begin{aligned} \mathbf{0} &= \begin{bmatrix} -\mathbf{B}\boldsymbol{\Lambda} & \mathbf{S}\boldsymbol{\Gamma} \end{bmatrix} \mathbf{C}, \\ &= \begin{bmatrix} 2\lambda_1 & 0 & -\gamma_1 & -\gamma_2 & \gamma_3 & \gamma_4 \\ 0 & 2\lambda_2 & \gamma_1 & -\gamma_2 & -\gamma_3 & \gamma_4 \end{bmatrix} \begin{bmatrix} -1 & 0 & 1 & 0 \\ 0 & -1 & 0 & 1 \\ -1 & 1 & 0 & 0 \\ 0 & -1 & 1 & 0 \\ 0 & 0 & -1 & 1 \\ 1 & 0 & 0 & -1 \end{bmatrix}, \\ &= \begin{bmatrix} \gamma_1 + \gamma_4 - 2\lambda_1 & \gamma_2 - \gamma_1 & 2\lambda_1 - \gamma_2 - \gamma_3 & \gamma_3 - \gamma_4 \\ \gamma_4 - \gamma_1 & \gamma_1 + \gamma_2 - 2\lambda_2 & \gamma_3 - \gamma_2 & 2\lambda_2 - \gamma_3 - \gamma_4 \end{bmatrix}. \end{aligned}$$

This requires

$$\lambda_1 = \lambda_2 = \gamma_1 = \gamma_2 = \gamma_3 = \gamma_4 = \gamma \geq 0.$$

Note that the above conditions involve six variables and eight equations, which require the solution of an under-determined system of linear equations on Λ and Γ. If we drop one column of $\mathbf{C}$, this reduces the number of equations to six, which produces a square and nonsingular system of equations to be solved.

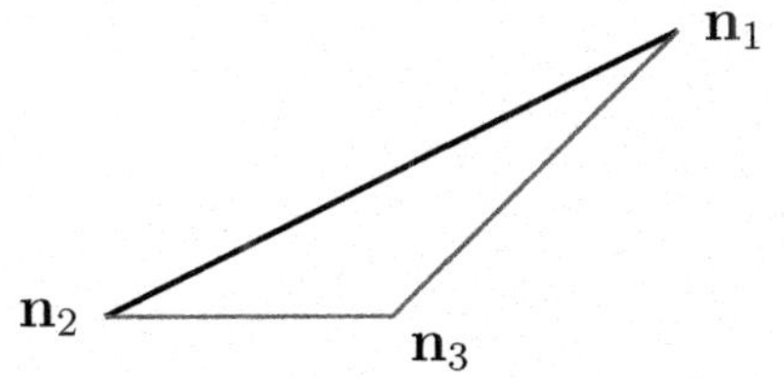

Figure 2.2: Simple planar structure

Example 2.2

Consider a structure with three ($n = 3$) nodes and three ($m = 3$) members, one bar and two strings

$$\mathbf{b}_1 = \mathbf{n}_2 - \mathbf{n}_1, \qquad \mathbf{s}_1 = \mathbf{n}_3 - \mathbf{n}_2, \qquad \mathbf{s}_2 = \mathbf{n}_1 - \mathbf{n}_3,$$

as shown in Figure 2.2. Let

$$\mathbf{N} = \begin{bmatrix}\mathbf{n}_1 & \mathbf{n}_3 & \mathbf{n}_3\end{bmatrix}, \qquad \mathbf{B} = \mathbf{b}_1, \qquad \mathbf{S} = \begin{bmatrix}\mathbf{s}_1 & \mathbf{s}_2\end{bmatrix}, \qquad \mathbf{M} = \begin{bmatrix}\mathbf{B} & \mathbf{S}\end{bmatrix}.$$

The connectivity matrix is

$$\mathbf{C} = \begin{bmatrix} -1 & 1 & 0 \\ 0 & -1 & 1 \\ 1 & 0 & -1 \end{bmatrix} \in \mathbb{R}^{3\times 3},$$

and the structure will be in equilibrium if

$$\mathbf{C}^T \mathbf{\Sigma} \mathbf{M}^T = \mathbf{0}.$$

This equilibrium is nontrivial if $\Sigma\mathbf{M}$ is not null and belongs in the null space of $\mathbf{C}^T$. The null space of $\mathbf{C}^T$ is composed of scalar multiples of the vector $\mathbf{1}_3$, so that

$$\mathbf{\Sigma}\mathbf{M}^T = \mathbf{1}_3 \mathbf{z}^T, \qquad \mathbf{z} \in \mathbb{R}^3.$$

In other words $\sigma_i \mathbf{m}_i = \mathbf{z}$, for all $i = 1, \ldots, 3$, i.e., all members should be aligned with the vector $\mathbf{z}$. Furthermore, the two strings must be pointing in the same direction for $\sigma_2 > 0$, $\sigma_3 > 0$, which implies that

$$\mathbf{n}_3 = \alpha \mathbf{n}_1 + (1 - \alpha)\mathbf{n}_2,$$

for some $0 \leq \alpha \leq 1$. Hence

$$\begin{aligned}
\mathbf{M} &= \begin{bmatrix}\mathbf{n}_2 - \mathbf{n}_1 & \mathbf{n}_3 - \mathbf{n}_2 & \mathbf{n}_1 - \mathbf{n}_3\end{bmatrix}, \\
&= \begin{bmatrix}\mathbf{n}_2 - \mathbf{n}_1 & \alpha\mathbf{n}_1 - \alpha\mathbf{n}_2 & (1-\alpha)\mathbf{n}_1 - (1-\alpha)\mathbf{n}_2\end{bmatrix}, \\
&= \begin{bmatrix}\mathbf{n}_1 & \mathbf{n}_2\end{bmatrix} \begin{bmatrix} -1 & \alpha & (1-\alpha) \\ 1 & -\alpha & -(1-\alpha) \end{bmatrix},
\end{aligned}$$

and

$$\begin{aligned}\mathbf{M\Sigma C} &= \begin{bmatrix}\mathbf{n}_1 & \mathbf{n}_2\end{bmatrix}\begin{bmatrix}-\sigma_1 & \sigma_2\alpha & \sigma_3(1-\alpha)\\ \sigma_1 & -\sigma_2\alpha & -\sigma_3(1-\alpha)\end{bmatrix}\begin{bmatrix}-1 & 1 & 0\\ 0 & -1 & 1\\ 1 & 0 & -1\end{bmatrix},\\ &= \begin{bmatrix}\mathbf{n}_1 & \mathbf{n}_2\end{bmatrix}\begin{bmatrix}\sigma_3(1-\alpha)+\sigma_1 & -(\sigma_1+\sigma_2\alpha) & \sigma_2\alpha-\sigma_3(1-\alpha)\\ -[\sigma_3(1-\alpha)+\sigma_1] & \sigma_1+\sigma_2\alpha & -[\sigma_2\alpha-\sigma_3(1-\alpha)]\end{bmatrix},\end{aligned}$$

which implies

$$\sigma_2 = -\frac{\sigma_1}{\alpha}, \qquad \sigma_3 = -\frac{\sigma_1}{1-\alpha}.$$

2.4.1 Affine Transformations

The general equilibrium conditions for a structure can be written as the matrix equation

$$\mathbf{NC}^T\mathbf{\Sigma C} = \mathbf{0}.$$

The above condition remains valid even when $\mathbf{N}$ and consequently $\mathbf{M}$ are subject to a *nonsingular affine node transformation* T. Such transformations are characterized by a nonsingular otherwise arbitrary matrix $\mathbf{T} \in \mathbb{R}^{3\times 3}$ and a vector $\mathbf{t} \in \mathbb{R}^3$ in the form

$$T : \mathbb{R}^{3\times n} \to \mathbb{R}^{3\times n}, \quad T(\mathbf{N}) := \mathbf{TN} + \mathbf{t}\mathbf{1}_n^T.$$

As shown at the end of the chapter, the equilibrium remains unaffected by such transformations. Affine node transformations do not change $\mathbf{\Sigma}$ or its sign, which will therefore imply that a tensegrity structure in equilibrium, for which $\Gamma \succ 0$, will remain a feasible tensegrity structure under such transformations. Also note that the matrix $\mathbf{\Sigma}$ remains invariant under such transformations but not $\mathbf{M}$. In particular, if T does not preserve the length of the members in $\mathbf{M}$, then the force will also be modified. For instance, a member $\mathbf{m}$ will be transformed to $\mathbf{Tm}$ and

$$\sigma_k(\|\mathbf{m}_k\|) = \sigma_k(\|\mathbf{Tm}_k\|) = \sigma_k$$

will remain the same but

$$\frac{V_k'(\|\mathbf{Tm}_k\|)}{V_k'(\|\mathbf{m}_k\|)} = \frac{\|\mathbf{Tm}_k\|}{\|\mathbf{m}_k\|}$$

will be different from one if $\|\mathbf{m}_k\| \neq \|\mathbf{Tm}_k\|$.

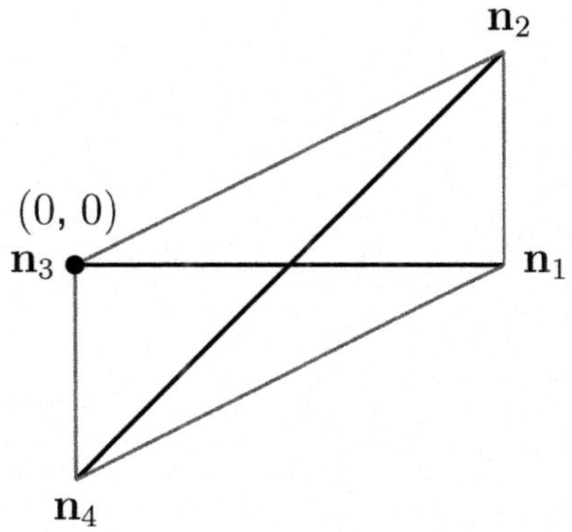

Figure 2.3: Planar tensegrity cross subject to affine node transformation

Example 2.3

We want to find out whether the planar tensegrity cross structure considered in previous examples can be put in equilibrium if the nodes are now located at

$$\tilde{\mathbf{N}} = \begin{bmatrix} 2 & 2 & 0 & 0 \\ 1 & 2 & 1 & 0 \end{bmatrix},$$

as shown in Figure 2.3. By noting that

$$\tilde{\mathbf{N}} = T(\mathbf{N})$$

for the affine transformation defined by

$$\mathbf{T} = \begin{bmatrix} 1 & 1 \\ 0 & 1 \end{bmatrix}, \qquad \mathrm{t} = \begin{pmatrix} 1 \\ 0 \end{pmatrix},$$

one can immediately answer yes. Note, however, that this transformation affects the length of the members and, consequently, the magnitude of the forces

$$V_k'(\|\mathbf{Tm}_k\|) = \begin{bmatrix} 2 & 2\sqrt{2} & 1 & \sqrt{5} & 1 & \sqrt{5} \end{bmatrix}.$$

2.4.2 Dual Structures

Another operation that preserves equilibrium is load reversal. Indeed, if

$$\mathbf{NC}^T\mathbf{\Sigma C} = \mathbf{0},$$

then trivially

$$\mathbf{NC}^T(-\mathbf{\Sigma})\mathbf{C} = \mathbf{0}.$$

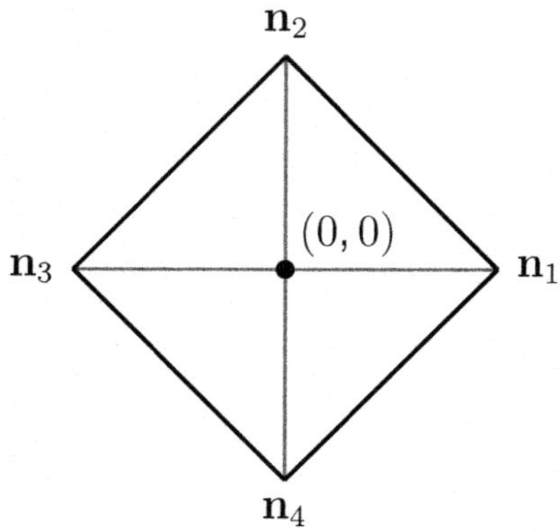

(a) Tensegrity cross

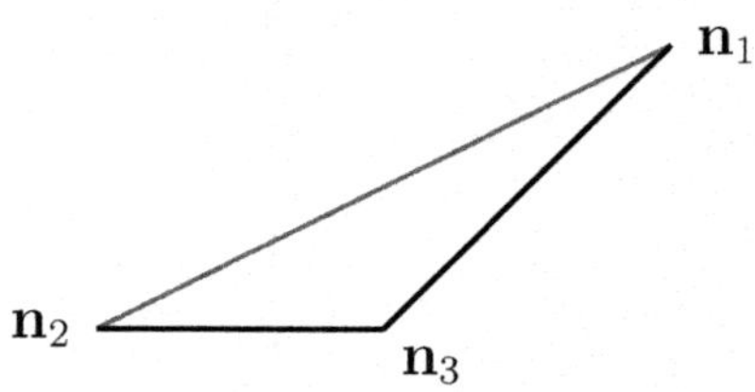

(b) Simple structure

Figure 2.4: Dual structures

This simple observation will produce interesting structures, combining primals and duals to produce yet another tensegrity structure preserving equilibrium. Of course, in the dual transformation bars become strings and strings become bars.

Example 2.4

Examples of dual structures are given in Figure 2.4, where 2.4(a) is the dual tensegrity cross in Figure 2.1 and 2.4(b) is the dual of the simple planar structure in Figure 2.2.

2.4.3 Class 1 Tensegrity Structures

A class one tensegrity structure with m_b bars have exactly $n = 2\,m_b$ nodes. Furthermore, each node belongs in exactly one bar. This one-to-one correspondence between nodes and bar nodes lets us formulate the equilibrium analysis in a particularly interesting set of coordinates.

Partition the connectivity matrix $\mathbf{C}$ as

$$\mathbf{C} = \begin{bmatrix} \mathbf{C}_B \\ \mathbf{C}_S \end{bmatrix},$$

such that

$$\mathbf{M} = \begin{bmatrix} \mathbf{B} & \mathbf{S} \end{bmatrix} = \mathbf{N} \begin{bmatrix} \mathbf{C}_B^T & \mathbf{C}_S^T \end{bmatrix}.$$

Because any node in the structure is a bar node, the node matrix can always be arranged so that

$$\mathbf{C}_B = \begin{bmatrix} -\mathbf{I} & \mathbf{I} \end{bmatrix}, \qquad \mathbf{C}_S = \begin{bmatrix} \mathbf{C}_{S1} & \mathbf{C}_{S2} \end{bmatrix}.$$

As shown at the end of the chapter, the equilibrium conditions for this particular set of coordinates is of the form

$$\mathbf{B}\boldsymbol{\Lambda} = \mathbf{S}\boldsymbol{\Gamma}\mathbf{C}_{S1} = -\mathbf{S}\boldsymbol{\Gamma}\mathbf{C}_{S2}, \qquad \boldsymbol{\Gamma} \succeq 0,$$

which can be given the following interesting interpretation: the sum of the string forces on node "1" of all bars, $\mathbf{S}\boldsymbol{\Gamma}\mathbf{C}_{S1}$, and the sum of the string forces on node "2" of all bars, $\mathbf{S}\boldsymbol{\Gamma}\mathbf{C}_{S2}$, must be collinear with the bar vectors and equal, with opposite signs, to the bar forces, $\mathbf{B}\boldsymbol{\Lambda}$.

Example 2.5

Consider the planar tensegrity cross structure from Figure 2.1. The string connectivity matrix can be partitioned as

$$\mathbf{C}_S = \begin{bmatrix} \mathbf{C}_{S1} & \mathbf{C}_{S2} \end{bmatrix} \in \mathbb{R}^{4\times 4},$$

where

$$\mathbf{C}_{S1} = \begin{bmatrix} -1 & 1 \\ 0 & -1 \\ 0 & 0 \\ 1 & 0 \end{bmatrix}, \qquad \mathbf{C}_{S2} = \begin{bmatrix} 0 & 0 \\ 1 & 0 \\ -1 & 1 \\ 0 & -1 \end{bmatrix}.$$

Therefore,

$$\mathbf{S}\boldsymbol{\Gamma}\mathbf{C}_{S1} = \begin{bmatrix} \gamma_1+\gamma_4 & \gamma_2-\gamma_1 \\ \gamma_4-\gamma_1 & \gamma_1+\gamma_2 \end{bmatrix},$$

$$\mathbf{S}\boldsymbol{\Gamma}\mathbf{C}_{S2} = \begin{bmatrix} -(\gamma_2+\gamma_3) & \gamma_3-\gamma_4 \\ \gamma_3-\gamma_2 & -(\gamma_3+\gamma_4) \end{bmatrix},$$

$$\mathbf{B}\boldsymbol{\Lambda} = \begin{bmatrix} -2\lambda_1 & 0 \\ 0 & -2\lambda_2 \end{bmatrix}.$$

The equilibrium conditions

$$\mathbf{B}\boldsymbol{\Lambda} = \mathbf{S}\boldsymbol{\Gamma}\mathbf{C}_{S1} = -\mathbf{S}\boldsymbol{\Gamma}\mathbf{C}_{S2}, \qquad \boldsymbol{\Gamma} \succeq 0,$$

imply

$$\lambda_1 = \lambda_2 = \gamma_1 = \gamma_2 = \gamma_3 = \gamma_4 = \gamma \geq 0.$$

2.5 Stiffness Matrix

The stiffness matrix $\mathbf{K}$ can be computed as

$$\mathbf{K}(\mathbf{n}) = -\partial_{\mathbf{n}}\mathbf{f}(\mathbf{n}).$$

We show at the end of the chapter that

$$\mathbf{K}(\mathbf{n}) = \sum_{k=1}^{m} \mathbf{K}_k(\mathbf{n}), \qquad \mathbf{K}_k(\mathbf{n}) = \mathbf{d}_k \mathbf{d}_k^T \otimes \mathbf{L}_k(\mathbf{m}_k),$$

where

$$\mathbf{L}_k(\mathbf{m}_k) = \sigma_k(\|\mathbf{m}_k\|) \left[\mathbf{I}_3 - \frac{\mathbf{m}_k \mathbf{m}_k^T}{\|\mathbf{m}_k\|^2} \right] + V_k''(\|\mathbf{m}_k\|) \frac{\mathbf{m}_k \mathbf{m}_k^T}{\|\mathbf{m}_k\|^2}.$$

Using the fact that $(\mathbf{x}\mathbf{y}^T \otimes \mathbf{A}) = (\mathbf{x} \otimes \mathbf{I})\mathbf{A}(\mathbf{y}^T \otimes \mathbf{I})$ we can rewrite the stiffness matrix as a matrix product of the form

$$\mathbf{K}(\mathbf{n}) = (\mathbf{C}^T \otimes \mathbf{I}_3) \operatorname{diag}[\mathbf{L}_1(\mathbf{m}_1), \ldots, \mathbf{L}_m(\mathbf{m}_m)](\mathbf{C} \otimes \mathbf{I}_3).$$

One may also find it useful to split $\mathbf{K}(\mathbf{n})$ into two components,

$$\mathbf{K}(\mathbf{n}) = \mathbf{K}_\sigma(\mathbf{n}) + \mathbf{K}_\phi(\mathbf{n}),$$

where

$$\mathbf{K}_\sigma(\mathbf{n}) := \mathbf{C}^T \mathbf{\Sigma} \mathbf{C} \otimes \mathbf{I}_3, \qquad \mathbf{K}_\phi(\mathbf{n}) := (\mathbf{C}^T \otimes \mathbf{I}_3)\, \mathbf{\Phi}\, (\mathbf{C} \otimes \mathbf{I}_3),$$

with

$$\mathbf{\Phi} := \operatorname{diag} \left\{ [V_1''(\|\mathbf{m}_1\|) - \sigma_1(\|\mathbf{m}_1\|)] \frac{\mathbf{m}_1 \mathbf{m}_1^T}{\|\mathbf{m}_1\|^2}, \ldots, \right.$$
$$\left. [V_m''(\|\mathbf{m}_m\|) - \sigma_m(\|\mathbf{m}_m\|)] \frac{\mathbf{m}_m \mathbf{m}_m^T}{\|\mathbf{m}_m\|^2} \right\}.$$

The first component $\mathbf{K}_\sigma$ results mostly from *prestress* while the second component $\mathbf{K}_\phi$ results mostly from *material*. Indeed, in equilibrium with no prestress, that is, $\mathbf{\Sigma} = 0$, only the second component is present. Furthermore, matrix $\mathbf{\Phi}$ is directly related to material properties. For instance, if forces are coming from linear elastic elements and $\mathbf{\Sigma} = 0$ then the scalar coefficients in matrix $\mathbf{\Phi}$ is the material stiffness $V_k''(\|\mathbf{m}_k\|) = \kappa_k$.

Example 2.6

For the planar tensegrity cross structure considered in the previous example assume we have elastic bars with stiffness κ_B and elastic strings with stiffness κ_S. In this case, one can compute for the two bars

$$\mathbf{L}_1 = \operatorname{diag}(\kappa_B, -\gamma), \qquad \mathbf{L}_2 = \operatorname{diag}(-\gamma, \kappa_B),$$

and for the four strings

$$\mathbf{L}_i = \frac{1}{2} \begin{bmatrix} \kappa_S + \gamma & -(\kappa_S - \gamma) \\ -(\kappa_S - \gamma) & \kappa_S + \gamma \end{bmatrix}, \qquad i = \{3, 5\},$$

$$\mathbf{L}_i = \frac{1}{2} \begin{bmatrix} \kappa_S + \gamma & \kappa_S - \gamma \\ \kappa_S - \gamma & \kappa_S + \gamma \end{bmatrix}, \qquad i = \{4, 6\}.$$

The entire matrix $\mathbf{K}$ is constructed as

$$\mathbf{K} = \begin{bmatrix} \mathbf{L}_1 + \mathbf{L}_3 + \mathbf{L}_6 & -\mathbf{L}_3 & -\mathbf{L}_1 & -\mathbf{L}_6 \\ -\mathbf{L}_3 & \mathbf{L}_2 + \mathbf{L}_3 + \mathbf{L}_4 & -\mathbf{L}_4 & -\mathbf{L}_2 \\ -\mathbf{L}_1 & -\mathbf{L}_4 & \mathbf{L}_1 + \mathbf{L}_4 + \mathbf{L}_5 & -\mathbf{L}_5 \\ -\mathbf{L}_6 & -\mathbf{L}_2 & -\mathbf{L}_5 & \mathbf{L}_2 + \mathbf{L}_5 + \mathbf{L}_6 \end{bmatrix}.$$

Example 2.7

For the planar simple structure in Figure 2.2 assume we have elastic bars with stiffness κ_B and elastic strings with stiffness κ_S. Let

$$\mathbf{n}_1 = \mathbf{e}_x, \qquad \mathbf{n}_2 = -\mathbf{e}_x, \qquad \sigma_1 = -\gamma, \qquad \sigma_2 = \gamma/\alpha, \qquad \sigma_3 = \gamma/(1-\alpha),$$

for some $\gamma > 0$. Compute

$$\mathbf{L}_1 = \operatorname{diag}(\kappa_B, -\gamma), \quad \mathbf{L}_2 = \operatorname{diag}(\kappa_S, \gamma/\alpha), \quad \mathbf{L}_3 = \operatorname{diag}(\kappa_S, \gamma/(1-\alpha)).$$

Matrix $\mathbf{K}$ is

$$\mathbf{K} = \begin{bmatrix} \mathbf{L}_1 + \mathbf{L}_3 & -\mathbf{L}_1 & -\mathbf{L}_3 \\ -\mathbf{L}_1 & \mathbf{L}_1 + \mathbf{L}_2 & -\mathbf{L}_2 \\ -\mathbf{L}_3 & -\mathbf{L}_2 & \mathbf{L}_2 + \mathbf{L}_3 \end{bmatrix}.$$

2.5.1 Modes and Modal Vectors

The eigenvalues of the matrix $\mathbf{K}$ are called *modes* of the structure and the associated eigenvectors, *modal vectors*. Modes and modal vectors satisfy the relationship

$$\mathbf{K}\,\mathbf{h}_\xi = \xi\,\mathbf{h}_\xi, \qquad \mathbf{h}_\xi \neq \mathbf{0}.$$

An interpretation of modes and modal vectors can be given by looking at the potential function around an equilibrium point $\mathbf{n}$. On a neighborhood of the equilibrium we can approximate V by its second-order Taylor expansion. That is, for small values of $\epsilon > 0$ we shall have

$$V(\mathbf{n} + \epsilon\mathbf{h}) \approx V(\mathbf{n}) + \frac{1}{2}\epsilon^2 \mathbf{h}^T \mathbf{K}\mathbf{h}.$$

If the nodes are displaced in the direction $\mathbf{h}$, thus changing the potential, the structure generates internal forces which can be approximated by

$$\mathbf{f}(\epsilon\mathbf{h}) = -\partial_{\mathbf{h}} V(\mathbf{n} + \epsilon\mathbf{h}) \approx -\epsilon\mathbf{K}\mathbf{h}.$$

If the displacement $\mathbf{h}$ is in the direction of a modal vector, i.e., $\mathbf{h} = \mathbf{h}_\xi$ then

$$\mathbf{f}(\epsilon\mathbf{h}_\xi) \approx -\epsilon\mathbf{K}\mathbf{h}_\xi = -\xi\epsilon\mathbf{h}_\xi.$$

That is, the force generated by the structure in response to a nodal displacement in a modal vector direction has magnitude $|\xi|$ and is on the opposite direction of the displacement $\mathbf{h}_\xi$ when ξ is positive.

As we shall see soon, in a stable structure, all modes shall be nonnegative, and most of them should be positive. However, at least six eigenvalues of $\mathbf{K}$ will always be zero. These eigenvalues are called *rigid body modes.* These modes are called rigid because they represent displacements, rigid translations and rigid rotations, that will not elongate any member of the structure, thus keeping the potential function V constant. As shown at the end of the chapter, rigid body modes are associated with null eigenvalues of matrix $\mathbf{K}$. The remaining eigenvalues are called *elastic modes.*

Example 2.8

The eigenvalues (modes) of $\mathbf{K}$ can be computed symbolically for the planar cross tensegrity structure in the form

$$0,\quad 0,\quad 0,\quad 2\left(\kappa_B+\gamma\right),\quad 2\left(\kappa_S-\gamma\right),\quad 2\left(\kappa_S+\gamma\right),\quad 2\left(\kappa_S+\gamma\right),\quad 2\left(\kappa_B+\kappa_S\right),$$

and the associated eigenvectors (modal vectors) are the columns of the matrix

$$\mathbf{H}_\xi = \begin{bmatrix} -1 & 1 & 1 & 1 & 0 & 0 & -1 & -1 \\ 2 & 0 & -1 & 0 & -1 & -1 & 0 & 0 \\ -2 & 1 & 2 & 0 & -1 & 0 & 1 & 0 \\ 1 & 0 & 0 & -1 & 0 & 1 & 0 & -1 \\ -1 & 1 & 1 & -1 & 0 & 0 & -1 & 1 \\ 0 & 0 & 1 & 0 & 1 & -1 & 0 & 0 \\ 0 & 1 & 0 & 0 & 1 & 0 & 1 & 0 \\ 1 & 0 & 0 & 1 & 0 & 1 & 0 & 1 \end{bmatrix}.$$

The first three eigenvalues are the rigid body modes, and the nodal displacements associated with the rigid modal vectors are illustrated in Figure 2.5. The second mode is a pure translation and the first and third modes are combinations of rotations and translations. The last five eigenvalues are elastic modes. The associated modal vector nodal displacements are shown in Figure 2.6.

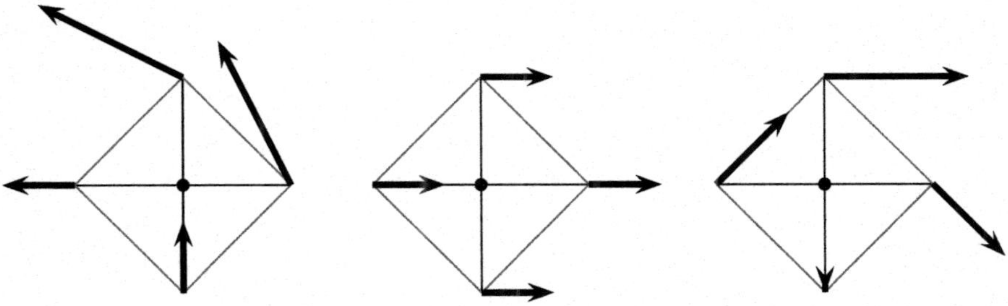

(a) Mode #1 (rigid) (b) Mode #2 (rigid) (c) Mode #3 (rigid)

Figure 2.5: Planar tensegrity cross rigid modal vectors

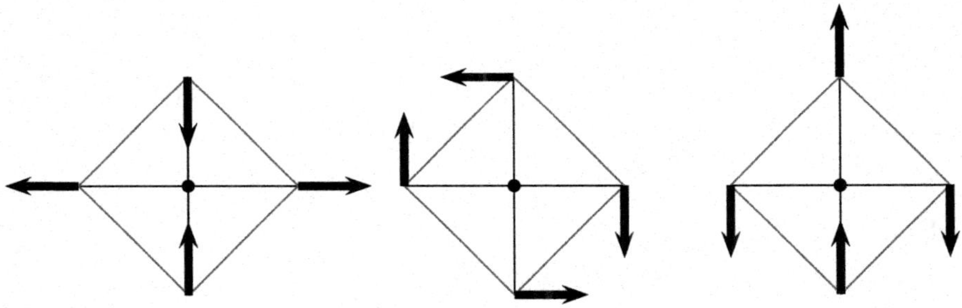

(a) Mode #4 (b) Mode #4 (c) Mode #5

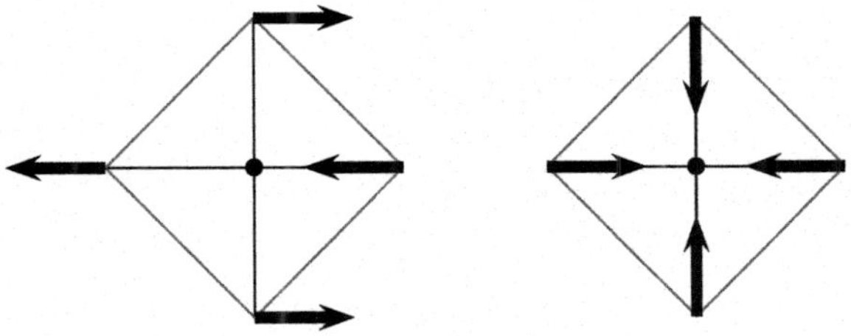

(d) Mode #6 (e) Mode #7

Figure 2.6: Planar tensegrity cross elastic modal vectors

Example 2.9

The eigenvalues (modes) of $\mathbf{K}$ can be computed symbolically for the planar simple structure in the form

$$0, \quad 0, \quad 0, \quad 2\left[\alpha^{-1}(1-\alpha)^{-1}-1\right]\gamma, \quad 3\kappa_S, \quad 2\kappa_B+\kappa_S,$$

and the associated eigenvectors (modal vectors) are the columns of the matrix

$$\mathbf{H}_\xi = \begin{bmatrix} 0 & 1 & 0 & 0 & -1/2 & -1 \\ \alpha^{-1} & 0 & -\alpha^{-1}(1-\alpha) & -\alpha & 0 & 0 \\ 0 & 1 & 0 & 0 & -1/2 & 1 \\ 0 & 0 & 1 & -(1-\alpha) & 0 & 0 \\ 0 & 1 & 0 & 0 & 1 & 0 \\ 1 & 0 & 0 & 1 & 0 & 0 \end{bmatrix}.$$

The first three eigenvalues are the rigid body modes.

2.5.2 Eliminating Rigid Body Modes

It is often useful to eliminate rigid body modes from the stiffness matrix, hence producing information on the structure's response in directions that do produce member deformation. What is needed to eliminate rigid body modes from the stiffness matrix is to build a basis for all rigid modal vectors $\mathbf{H}_t$ and $\mathbf{H}_r$. One such choice for a basis is given by

$$\mathbf{h}_R = \mathbf{B}_R \mathbf{z}_R, \qquad \mathbf{z}_R \in \mathbb{R}^6,$$

where

$$\mathbf{B}_R = \begin{bmatrix} \mathbf{1}_n \otimes \mathbf{I}_3 & \text{vec}[\text{skew}(\mathbf{e}_x)\mathbf{N}] & \text{vec}[\text{skew}(\mathbf{e}_y)\mathbf{N}] & \text{vec}[\text{skew}(\mathbf{e}_z)\mathbf{N}] \end{bmatrix}.$$

Then construct an orthonormal basis for all directions $\mathbf{h}_e$ such that

$$\mathbf{h}_E^T \mathbf{h}_R = \mathbf{0},$$

that is, an orthonormal basis for the null space of the subspace spanned by $\mathbf{h}_R$. All such vectors can be parametrized as

$$\mathbf{h}_E = \mathbf{B}_E \mathbf{z}_E, \qquad \mathbf{z}_E \in \mathbb{R}^{3n-6}, \qquad \mathbf{B}_E^T \mathbf{B}_E = \mathbf{I}_{3n-6}.$$

In fact, any displacement direction $\mathbf{h}$ can be decomposed as

$$\mathbf{h} = \mathbf{B}_R \mathbf{z}_R + \mathbf{B}_E \mathbf{z}_E.$$

Moreover, around an equilibrium point $\mathbf{n}$

$$\begin{aligned} V(\mathbf{n} + \epsilon \mathbf{h}) - V(\mathbf{n}) &= V(\mathbf{n} + \epsilon[\mathbf{B}_R \mathbf{z}_R + \mathbf{B}_E \mathbf{z}_E]) - V(\mathbf{n}) \\ &= \frac{1}{2}\epsilon^2 \left(\mathbf{B}_R \mathbf{z}_R + \mathbf{B}_E \mathbf{z}_E\right)^T \mathbf{K} \left(\mathbf{B}_R \mathbf{z}_R + \mathbf{B}_E \mathbf{z}_E\right) + O(\epsilon^3) \\ &= \frac{1}{2}\epsilon^2 \mathbf{z}_E^T \mathbf{B}_E^T \mathbf{K} \mathbf{B}_E \mathbf{z}_E + O(\epsilon^3). \end{aligned}$$

The matrix

$$\mathbf{K}_E = \mathbf{B}_E^T \mathbf{K} \mathbf{B}_E$$

is the elastic component of the stiffness matrix from which the rigid body modes have been eliminated.

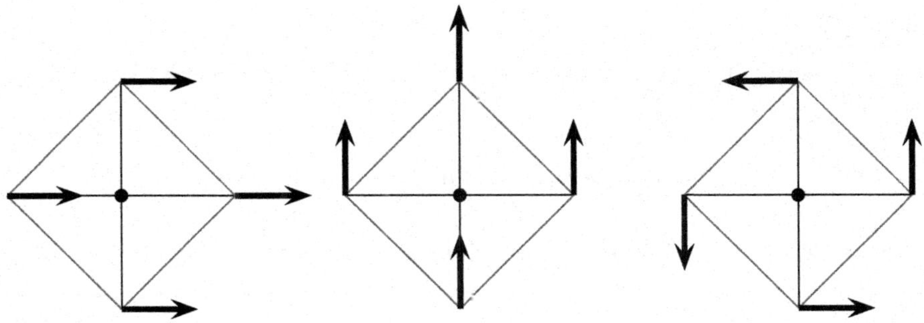

(a) Mode #1 (rigid) (b) Mode #2 (rigid) (c) Mode #3 (rigid)

Figure 2.7: Planar tensegrity cross rigid modal vectors. Decoupled translations and rotation

Example 2.10

First note that $\mathbf{B}_R$ can be expressed as a function of the eigenvectors depicted in Figure 2.5 in the form

$$\mathbf{B}_R = \begin{bmatrix}\mathbf{h}_2 & (\mathbf{h}_1 + \mathbf{h}_3) & (\mathbf{h}_2 - \mathbf{h}_3)\end{bmatrix}.$$

The nodal displacements resulting from $\mathbf{B}_R$ are shown in Figure 2.7. In this figure we see that the translations and rotations are now decoupled and orthogonal. Now compute an orthonormal basis for the null space of $\mathbf{B}_R$, such as

$$\mathbf{B}_E = \frac{1}{2}\begin{bmatrix} 1 & -\sqrt{15}/3 & 0 & -\sqrt{3}/3 & 0 \\ -1 & -\sqrt{15}/5 & -\sqrt{15}/15 & 0 & -\sqrt{3}/3 \\ -1 & \sqrt{15}/15 & -\sqrt{15}/5 & -\sqrt{3}/3 & 0 \\ 0 & 0 & 0 & 0 & \sqrt{3} \\ 0 & 0 & 0 & \sqrt{3} & 0 \\ 0 & 0 & \sqrt{15}/3 & 0 & -\sqrt{3}/3 \\ 0 & 4\sqrt{15}/15 & \sqrt{15}/5 & -\sqrt{3}/3 & 0 \\ 1 & \sqrt{15}/5 & -4\sqrt{15}/15 & 0 & -\sqrt{3}/3 \end{bmatrix}.$$

2.5.3 Stability

If at the equilibrium point $\mathbf{n}$ the elastic stiffness matrix is positive definite, that is,

$$\mathbf{K}_E = \mathbf{B}_E^T \mathbf{K} \mathbf{B}_E \succ 0,$$

then the equilibrium $\mathbf{n}$ is said to be *stable*. The idea is that the potential function must increase for any small displacements in all elastic directions $\mathbf{h}_E = \mathbf{B}_E \mathbf{z}_E$, thus generating a restoring force that tends to bring the structure back to its original equilibrium in an open neighborhood of $\mathbf{n}$.

Example 2.11

The tensegrity planar cross is on a stable equilibrium since $\gamma > 0$ and the smallest eigenvalue of $\mathbf{K}_E$ is

$$\underline{\xi} = \min_i \xi_i(\mathbf{K}_E) = 2 \min\{\kappa_B + \gamma, \kappa_S - \gamma\},$$

and we have from (2.7) that for feasibility $\kappa_S > \gamma$ so that $\underline{\xi} > 0$, hence $\mathbf{K}_E \succ 0$.

Example 2.12

The analysis of the planar simple structure in Figure 2.2 is a bit more subtle. First note that the smallest elastic eigenvalue of $\mathbf{K}_E$ is

$$\underline{\xi} = 2\left[\alpha^{-1}(1-\alpha)^{-1} - 1\right]\gamma,$$

whose value depends solely on the level of pretension γ and α, which tells the location of the node $\mathbf{n}_3$.

If $\gamma > 0$ then this eigenvalue is positive and the structure is stable. From the modal vector matrix $\mathbf{H}$, the displacements associated with this mode move all nodes perpendicular to the member directions. On a stable structure, the structure produces a force that brings the nodes back to the original equilibrium line.

If $\gamma = 0$ then this mode is null, which indicates that a *mechanism* exists in the structure, in this case an *infinitesimal mechanism* (see [CP91] for details). Nodes can move perpendicular to the members and the structure offers no resistance.

If $\gamma < 0$, in which case bar and strings reverse roles and we have the dual structure 2.4(b), the minimum eigenvalue is negative, which means that the structure is no longer stable. The structure produces a force that takes the nodes away from the original equilibrium line.

2.5.4 Eliminating Internal Nodes

In a tensegrity structure many nodes are internal to the structure and never interact with external forces. It may be useful to account for that effect while computing the stiffness matrix.

Let $\mathbf{n}$ be a configuration in equilibrium. Then the force generated by a small displacement of the nodes $\mathbf{n} + \mathbf{h}$ is approximately given by

$$\mathbf{f} \approx -\mathbf{K}\mathbf{h}.$$

Conversely, if an external force $\mathbf{f}$ is applied to the structure, the above equation predicts the internal displacements $\mathbf{h}$.

Now rearrange and partition the displacement and force vectors in the form

$$\mathbf{h} = \begin{bmatrix} \mathbf{h}^e \\ \mathbf{h}^i \end{bmatrix}, \qquad \mathbf{f} = \begin{bmatrix} \mathbf{f}^e \\ \mathbf{f}^i \end{bmatrix},$$

where $\mathbf{h}^i$ $(\mathbf{f}^i) \in \mathbf{R}^{3n^i}$ denote all displacements (forces) internal to the structure. Partitioning,

$$\mathbf{K} = \begin{bmatrix} \mathbf{K}_{ee} & \mathbf{K}_{ei}^T \\ \mathbf{K}_{ei} & \mathbf{K}_{ii} \end{bmatrix},$$

accordingly. The condition that a node is internal can be translated as

$$\mathbf{f}^i = \begin{bmatrix} \mathbf{K}_{ei} & \mathbf{K}_{ii} \end{bmatrix} \begin{bmatrix} \mathbf{h}^e \\ \mathbf{h}^i \end{bmatrix} = \mathbf{0}, \tag{2.9}$$

that is, that the external force on an internal node will be zero. Assume[1] that $\mathbf{K}_{ii} \succ 0$, in which case, all solutions to the above equation are given by

$$\mathbf{h} = \begin{bmatrix} \mathbf{I} \\ -\mathbf{K}_{ii}^{-1}\mathbf{K}_{ei} \end{bmatrix} \mathbf{h}^e.$$

These are the global displacements when no external force is applied on the internal nodes $\mathbf{n}^i$. Using this as the displacements in the power series expansion of $V(\mathbf{n} + \epsilon\mathbf{h})$ we obtain the reduced stiffness matrix as

$$\mathbf{K}^e = \mathbf{K}_{ee} - \mathbf{K}_{ei}\mathbf{K}_{ii}^{-1}\mathbf{K}_{ei}^T.$$

Note that the above matrix still contains all rigid body modes, since any rigid displacement $\mathbf{h}_R$ satisfies (2.9) as well. The rigid body modes can be eliminated by applying the procedure we studied previously on the reduced stiffness matrix $\mathbf{K}^e$ instead of $\mathbf{K}$.

Example 2.13

Assume that node $\mathbf{n}_3$ in the planar simple structure in Figure 2.2 is an internal node. The reduced stiffness matrix is then

$$\mathbf{K}_e = \begin{bmatrix} \mathbf{L}_1 + \mathbf{L}_3 & -\mathbf{L}_1 \\ -\mathbf{L}_1 & \mathbf{L}_1 + \mathbf{L}_2 \end{bmatrix} - \begin{bmatrix} -\mathbf{L}_3 \\ -\mathbf{L}_2 \end{bmatrix} (\mathbf{L}_2 + \mathbf{L}_3)^{-1} \begin{bmatrix} -\mathbf{L}_3 & -\mathbf{L}_2 \end{bmatrix}.$$

The eigenvalues (modes) of $\mathbf{K}_e$ can be computed symbolically in this simple case:

$$0, \quad 0, \quad 0, \quad 2\,\kappa_B + \kappa_S.$$

Note that all rigid body modes are still present in $\mathbf{K}^e$.

[1] If that is not the case, compute $\mathbf{h} = \begin{bmatrix} \mathbf{K}_{ei} & \mathbf{K}_{ii} \end{bmatrix}^{\perp} \mathbf{z}$.

2.6 External Forces

In a structure with n nodes let

$$\mathbf{w}_i \in \mathbb{R}^3, \quad i = 1, \dots, n,$$

be three-dimensional vectors describing external forces applied at each node of the structure. We define the external force vector and external force matrix as

$$\mathbf{w} = \begin{pmatrix} \mathbf{w}_1 \\ \vdots \\ \mathbf{w}_n \end{pmatrix} \in \mathbb{R}^{3n}, \qquad \mathbf{W} = \begin{bmatrix} \mathbf{w}_1 & \cdots & \mathbf{w}_n \end{bmatrix} \in \mathbb{R}^{3\times n},$$

respectively.

Equilibrium in the presence of external forces can be obtained by simply adding the linear potential to the total potential energy of the structure

$$V_{\mathbf{w}}(\mathbf{n}) = V(\mathbf{n}) + \mathbf{w}^T\mathbf{n},$$

or, in matrix form,

$$V_{\mathbf{W}}(\mathbf{N}) = V(\mathbf{N}) + \text{trace}(\mathbf{W}^T\mathbf{N}).$$

As in Section 2.4, an equilibrium point is a stationary point of the potential energy, hence

$$\partial_{\mathbf{N}} V_{\mathbf{W}}(\bar{\mathbf{N}}) = \partial_{\mathbf{N}} V(\bar{\mathbf{N}}) + \mathbf{W} = \mathbf{0}$$

here given directly in matrix form. In other words, a point $\bar{\mathbf{N}}$ in which the internal forces $\mathbf{F}$ balance the external forces $\mathbf{W}$, i.e.,

$$-\partial_{\mathbf{N}} V(\bar{\mathbf{n}}) = \mathbf{F}(\bar{\mathbf{N}}) = \mathbf{W}.$$

Most developments presented earlier in this chapter can be generalized to cope with external forces by simply adding the vector or matrix of external forces to the right-hand side of the equilibrium equations.

2.6.1 Optimal Volume of Loaded Structures

The derivations in this section follow mainly the ideas of A. G. M. Michell [Mic04].

Let $\bar{\sigma}_k$, $k = 1, \dots, m$, denote the *yield stress* of the material used to construct the kth member of a structure loaded with the external force $\mathbf{w}$. Assume that each member has a constant cross-section area A_k, $k = 1, \dots, m$. If such a structure is designed so that no members experience yielding then the maximum force density at any given member should be at most

$$\sigma_k(\|\mathbf{m}_k\|) \leq \bar{\sigma}_k \frac{A_k}{\|\mathbf{m}_k\|}, \qquad k = 1, \dots, m.$$

For minimal mass structures, equalities should hold on the above expressions. In this case, the volume of each member can be related to the force density through the formula

$$V_k = \|\mathbf{m}_k\| A_k = \frac{1}{\bar{\sigma}_k}\sigma_k(\|\mathbf{m}_k\|)\,\|\mathbf{m}_k\|^2, \qquad k = 1,\dots,m.$$

The total volume of a structure is then computed by the summation

$$V = \sum_{k=1}^{m} V_k = \sum_{k=1}^{m} \frac{1}{\bar{\sigma}_k}\sigma_k(\|\mathbf{m}_k\|)\,\|\mathbf{m}_k\|^2.$$

In a structure with m_b bars and m_s strings the above summation can be split into two parts:

$$V = V_b + V_s,$$
$$V_b := \sum_{k=1}^{m_b} \frac{1}{\bar{\lambda}_k}\lambda_k(\|\mathbf{b}_k\|)\,\|\mathbf{b}_k\|^2, \qquad V_s := \sum_{k=1}^{m_s} \frac{1}{\bar{\gamma}_k}\gamma_k(\|\mathbf{s}_k\|)\,\|\mathbf{s}_k\|^2.$$

Assume now that all bars are made of the same material, then

$$\bar{\lambda}_k = \bar{\lambda} \quad \text{for all } k = 1,\dots,m_b.$$

Likewise, for strings made of the same material

$$\bar{\gamma}_k = \bar{\gamma} \quad \text{for all } k = 1,\dots,m_s.$$

In this case the total volume is given by adding the components:

$$V_b := \frac{1}{\bar{\lambda}}\sum_{k=1}^{m_b} \lambda_k(\|\mathbf{b}_k\|)\|\mathbf{b}_k\|^2, \qquad V_s := \frac{1}{\bar{\gamma}}\sum_{k=1}^{m_s} \gamma_k(\|\mathbf{s}_k\|)\|\mathbf{s}_k\|^2.$$

In the next paragraphs we will show how to compute the volume using information from the equilibrium equations. We shall omit the dependence of the force densities λ_k and γ_k on the length of the kth member. Partitioning the members into bars and strings as in Section 2.4 we must have at equilibrium

$$\begin{bmatrix}-\mathbf{B}\mathbf{\Lambda} & \mathbf{S}\mathbf{\Gamma}\end{bmatrix}\mathbf{C} = \mathbf{W}, \qquad \mathbf{\Lambda} \succeq 0, \qquad \mathbf{\Gamma} \succeq 0,$$

where the positivity constraints ensure the bars and strings are carrying, respectively, compressive and tensile forces. Multiplication of this expression on the right by the transpose of the node matrix $\mathbf{N}$ produces

$$\begin{bmatrix}-\mathbf{B}\mathbf{\Lambda} & \mathbf{S}\mathbf{\Gamma}\end{bmatrix}\mathbf{C}\mathbf{N}^T = \mathbf{W}\mathbf{N}^T.$$

Recalling that

$$\mathbf{N}\mathbf{C}^T = \begin{bmatrix}\mathbf{B} & \mathbf{S}\end{bmatrix},$$

we obtain

$$\mathbf{S}\mathbf{\Gamma}\mathbf{S}^T - \mathbf{B}\mathbf{\Lambda}\mathbf{B}^T = \mathbf{W}\mathbf{N}^T.$$

Now take the trace of the above matrix on both sides

$$\text{trace}(\mathbf{S}\mathbf{\Gamma}\mathbf{S}^T) - \text{trace}(\mathbf{B}\mathbf{\Lambda}\mathbf{B}^T) = \text{trace}(\mathbf{W}\mathbf{N}^T).$$

Note that the traces on the left-hand side of the above expression also appear in the volume, which can be conveniently rewritten in the form

$$V = \frac{1}{\bar{\gamma}}\,\text{trace}(\mathbf{S}\mathbf{\Gamma}\mathbf{S}^T) + \frac{1}{\bar{\lambda}}\,\text{trace}(\mathbf{B}\mathbf{\Lambda}\mathbf{B}^T).$$

This observation is the key for the next result.

Assume that the compressive and tensile yield stresses are equal, i.e., $\bar{\gamma} = \bar{\lambda}$. Among all structures sharing the same set of external forces $\mathbf{w}_i$ applied at nodes $\mathbf{n}_i$ for $i = 1, \ldots, n$ the quantity J defined by

$$J = 2\,\bar{\lambda}\bar{\sigma}\,V + (\bar{\sigma} - \bar{\lambda})\,\text{trace}(\mathbf{W}\mathbf{N}^T)$$

is least when V itself is least. However, notice that

$$\begin{aligned} J &= 2\,\bar{\lambda}\bar{\gamma}\,V + (\bar{\gamma} - \bar{\lambda})\,\text{trace}(\mathbf{W}\mathbf{N}^T) \\ &= 2\,\bar{\lambda}\bar{\gamma}\left[\text{trace}(\mathbf{S}\mathbf{\Gamma}\mathbf{S}^T) + \text{trace}(\mathbf{B}\mathbf{\Lambda}\mathbf{B}^T)\right] \\ &\qquad + (\bar{\gamma} - \bar{\lambda})\left[\text{trace}(\mathbf{S}\mathbf{\Gamma}\mathbf{S}^T) - \text{trace}(\mathbf{B}\mathbf{\Lambda}\mathbf{B}^T)\right] \\ &= (\bar{\lambda} + \bar{\gamma})\left[\text{trace}(\mathbf{S}\mathbf{\Gamma}\mathbf{S}^T) + \text{trace}(\mathbf{B}\mathbf{\Lambda}\mathbf{B}^T)\right]. \end{aligned} \tag{2.10}$$

Minimization of J is therefore independent of the choice of material properties, $(\bar{\lambda} + \bar{\gamma})$, and depends only on the variable topology (lengths and orientation of members). We shall explore this extraordinary property first observed by Michell [Mic04]. Notice that, as pointed out by Rozvany [Roz96, Roz97], in the case when $\bar{\gamma} \neq \bar{\lambda}$, minimization of J may not lead to minimization of V, as claimed originally by Michell [Mic04]. This is the reason we limit our attention to the case $\bar{\gamma} = \bar{\lambda}$. We shall explore this extraordinary property at other points of this book.

2.7 Chapter Summary

The methods of this chapter form the foundation of work in all remaining chapters, where optimal designs will be derived to minimize mass or bound stiffness or optimize topology. There are many ways to connect strings and bars without obtaining a stable equilibrium, or without obtaining a structure stiff enough for engineering applications. This chapter provides the analytical tools to compute, or design for, stiffness, mass, and topology optimization later in the text. A network point of view is taken to characterize the structural forces and topologies. By defining a connectivity matrix, one can arrange any topology and efficiently determine stiffness, mass, and stability. A vector approach simplifies the equation structure and eliminates transcendental functions in the equations.

The form-finding equations are linear in the configuration variables. Hence the forces in all members are immediately known in closed form formulas, given the external forces and a specified configuration. Tools are also given to determine the modes of the structure of a given topology.

2.8 Advanced Material

2.8.1 Affine Transformations

That equilibrium remain unaffected by affine node transformations can be verified by writing

$$\begin{aligned} T(\mathbf{N})\mathbf{C}^T\mathbf{\Sigma}\mathbf{C} &= \left(\mathbf{TN} + \mathbf{t}\mathbf{1}_n^T\right)\mathbf{C}^T\mathbf{\Sigma}\mathbf{C}, \\ &= \mathbf{TNC}^T\mathbf{\Sigma}\mathbf{C} + \mathbf{t}\mathbf{1}_n^T\mathbf{C}^T\mathbf{\Sigma}\mathbf{C}, \\ &= \mathbf{TNC}^T\mathbf{\Sigma}\mathbf{C}, \\ &= \mathbf{0}, \end{aligned}$$

which is a consequence of the nonsingularity of $\mathbf{T}$ and the fact that $\mathbf{C}\mathbf{1}_n = \mathbf{0}$.

2.8.2 Class 1 Tensegrity Structures

Define the nonsingular matrix

$$\mathbf{T} = \begin{bmatrix} \mathbf{I} & \mathbf{I} \\ \mathbf{0} & \mathbf{I} \end{bmatrix},$$

for which

$$\mathbf{C}_B\mathbf{T} = \begin{bmatrix} -\mathbf{I} & \mathbf{0} \end{bmatrix}, \qquad \mathbf{C}_S\mathbf{T} = \begin{bmatrix} \mathbf{C}_{S1} & \mathbf{C}_{S1} + \mathbf{C}_{S2} \end{bmatrix},$$

and

$$\begin{aligned} &\mathbf{T}^T\left(-\mathbf{C}_B^T\mathbf{\Lambda}\mathbf{C}_B + \mathbf{C}_S^T\mathbf{\Gamma}\mathbf{C}_S\right)\mathbf{T} \\ &\quad = \begin{bmatrix} \mathbf{C}_{S1}^T\mathbf{\Gamma}\mathbf{C}_{S1} - \mathbf{\Lambda} & \mathbf{C}_{S1}^T\mathbf{\Gamma}(\mathbf{C}_{S1} + \mathbf{C}_{S2}) \\ (\mathbf{C}_{S1}^T + \mathbf{C}_{S2}^T)\mathbf{\Gamma}\mathbf{C}_{S1} & (\mathbf{C}_{S1}^T + \mathbf{C}_{S2}^T)\mathbf{\Gamma}(\mathbf{C}_{S1} + \mathbf{C}_{S2}) \end{bmatrix}. \end{aligned}$$

Define also the transformed nodal coordinates

$$\begin{bmatrix} \bar{\mathbf{N}}_1 & \bar{\mathbf{N}}_2 \end{bmatrix} = \bar{\mathbf{N}} = \mathbf{N}\mathbf{T}^{-T} = \begin{bmatrix} \mathbf{N}_1 & \mathbf{N}_2 \end{bmatrix} \begin{bmatrix} \mathbf{I} & \mathbf{0} \\ -\mathbf{I} & \mathbf{I} \end{bmatrix} = \begin{bmatrix} \mathbf{N}_1 - \mathbf{N}_2 & \mathbf{N}_2 \end{bmatrix},$$

and multiply the equilibrium conditions on the right by $\mathbf{T}$ to obtain the equivalent conditions

$$\begin{aligned}
\mathbf{0} &= \mathbf{N}\mathbf{C}^T\mathbf{\Sigma}\mathbf{C}\mathbf{T} \\
&= \bar{\mathbf{N}}\mathbf{T}^T\left(-\mathbf{C}_B^T\mathbf{\Lambda}\mathbf{C}_B + \mathbf{C}_S^T\mathbf{\Gamma}\mathbf{C}_S\right)\mathbf{T}, \\
&= \begin{bmatrix}\mathbf{N}_1 - \mathbf{N}_2 & \mathbf{N}_2\end{bmatrix}\begin{bmatrix}\mathbf{C}_{S1}^T\mathbf{\Gamma}\mathbf{C}_{S1} - \mathbf{\Lambda} & \mathbf{C}_{S1}^T\mathbf{\Gamma}(\mathbf{C}_{S1} + \mathbf{C}_{S2}) \\ (\mathbf{C}_{S1}^T + \mathbf{C}_{S2}^T)\mathbf{\Gamma}\mathbf{C}_{S1} & (\mathbf{C}_{S1}^T + \mathbf{C}_{S2}^T)\mathbf{\Gamma}(\mathbf{C}_{S1} + \mathbf{C}_{S2})\end{bmatrix}, \\
&= \left[(\mathbf{N}_1\mathbf{C}_{S1}^T + \mathbf{N}_2\mathbf{C}_{S2}^T)\mathbf{\Gamma}\mathbf{C}_{S1} - (\mathbf{N}_1 - \mathbf{N}_2)\mathbf{\Lambda} \quad (\mathbf{N}_1\mathbf{C}_{S1}^T + \mathbf{N}_2\mathbf{C}_{S2}^T)\mathbf{\Gamma}(\mathbf{C}_{S1} + \mathbf{C}_{S2})\right], \\
&= \left[\mathbf{N}\mathbf{C}_S^T\mathbf{\Gamma}\mathbf{C}_{S1} + \mathbf{N}\mathbf{C}_B^T\mathbf{\Lambda} \quad \mathbf{N}\mathbf{C}_S^T\mathbf{\Gamma}(\mathbf{C}_{S1} + \mathbf{C}_{S2})\right], \\
&= \left[\mathbf{S}\mathbf{\Gamma}\mathbf{C}_{S1} + \mathbf{B}\mathbf{\Lambda} \quad \mathbf{S}\mathbf{\Gamma}(\mathbf{C}_{S1} + \mathbf{C}_{S2})\right], \qquad \mathbf{\Lambda} \succeq 0.
\end{aligned}$$

2.8.3 Stiffness Matrix

Note that

$$-\partial_{\mathbf{n}_{i_k}}\mathbf{f}_k(\mathbf{n}) = [\mathbf{d}_k \otimes \partial_{\mathbf{n}_{i_k}}\sigma_k(\|\mathbf{m}_k\|)\,\mathbf{m}_k],$$

where

$$\begin{aligned}
\mathbf{L}_k(\mathbf{m}_k) &:= \partial_{\mathbf{n}_{i_k}}\sigma_k(\|\mathbf{m}_k\|)\mathbf{m}_k \\
&= \sigma_k(\|\mathbf{m}_k\|)\,\mathbf{I}_3 + \sigma_k'(\|\mathbf{m}_k\|)\,\mathbf{m}_k\,\frac{\mathbf{m}_k^T}{\|\mathbf{m}_k\|}, \\
&= \sigma_k(\|\mathbf{m}_k\|)\,\mathbf{I}_3 + \left[\frac{V_k''(\|\mathbf{m}_k\|)}{\|\mathbf{m}_k\|} - \frac{V_k'(\|\mathbf{m}_k\|)}{\|\mathbf{m}_k\|^2}\right]\frac{\mathbf{m}_k\mathbf{m}_k^T}{\|\mathbf{m}_k\|}, \\
&= \sigma_k(\|\mathbf{m}_k\|)\,\mathbf{I}_3 + [V_k''(\|\mathbf{m}_k\|) - \sigma_k(\|\mathbf{m}_k\|)]\,\frac{\mathbf{m}_k\mathbf{m}_k^T}{\|\mathbf{m}_k\|^2}, \\
&= \sigma_k(\|\mathbf{m}_k\|)\left[\mathbf{I}_3 - \frac{\mathbf{m}_k\mathbf{m}_k^T}{\|\mathbf{m}_k\|^2}\right] + V_k''(\|\mathbf{m}_k\|)\,\frac{\mathbf{m}_k\mathbf{m}_k^T}{\|\mathbf{m}_k\|^2}.
\end{aligned}$$

Because

$$\partial_{\mathbf{n}_{j_k}}\sigma_k(\|\mathbf{m}_k\|)\,\mathbf{m}_k = -\partial_{\mathbf{n}_{i_k}}\sigma_k(\|\mathbf{m}_k\|)\,\mathbf{m}_k = -\mathbf{L}_k(\mathbf{m}_k),$$

we have

$$\mathbf{K}(\mathbf{n}) = \sum_{k=1}^{m}\mathbf{K}_k(\mathbf{n}), \qquad \mathbf{K}_k(\mathbf{n}) = \mathbf{d}_k\mathbf{d}_k^T \otimes \mathbf{L}_k(\mathbf{m}_k).$$

2.8.4 Modes and Modal Vectors

To understand how rigid body modes are associated with null eigenvalues of matrix $\mathbf{K}$, assume, for simplicity, that the structure contains only linear elements. From (2.7) we must have

$$V_k''(\|\mathbf{m}_k\|) = \kappa_k > \sigma_k(\|\mathbf{m}_k\|),$$

so that $\mathbf{\Phi} \succeq 0$. Hence

$$\mathbf{K}_\phi \succeq 0.$$

Now define

$$\mathbf{H} = \begin{bmatrix} \mathbf{h}_1 & \cdots & \mathbf{h}_n \end{bmatrix} \in \mathbb{R}^{3\times n},$$

and compute

$$\begin{aligned} \mathrm{vec}(\mathbf{H})^T \mathbf{K}_\phi \, \mathrm{vec}(\mathbf{H}) &= \sum_{k=1}^{m} \frac{\kappa_k - \sigma_k(\|\mathbf{m}_k\|)}{\|\mathbf{m}_k\|^2} \mathrm{vec}(\mathbf{H})^T \left[\mathbf{d}_k \mathbf{d}_k^T \otimes \mathbf{m}_k \mathbf{m}_k^T\right] \mathrm{vec}(\mathbf{H}), \\ &= \sum_{k=1}^{m} \left[\kappa_k - \sigma_k(\|\mathbf{m}_k\|)\right] \left(\mathbf{d}_k^T \mathbf{H}^T \mathbf{m}_k / \|\mathbf{m}_k\|\right)^2 \geq 0. \end{aligned}$$

Note that the above expression is zero for some nonzero $\mathbf{H}$ if and only if

$$\mathbf{m}_k^T \mathbf{H} \mathbf{d}_k = \mathbf{0}, \quad \text{for all } k.$$

We will discuss two cases in which this is possible. The first is for all directions of the form

$$\mathbf{H}_t = \mathbf{v} \mathbf{1}_n^T,$$

where $\mathbf{v} \in \mathbb{R}^3$ is an arbitrary three-dimensional vector. This follows from the fact that $\mathbf{C}\mathbf{1}_n = \mathbf{0}$. The second case is when

$$\mathbf{H}_r = \mathrm{skew}(\mathbf{v})\,\mathbf{N},$$

again for an arbitrary $\mathbf{v} \in \mathbb{R}^3$. Indeed, for all k,

$$\begin{aligned} \mathbf{m}_k^T \mathbf{H}_r \mathbf{d}_k &= \mathbf{m}_k^T \,\mathrm{skew}(\mathbf{v})\,\mathbf{N}\mathbf{d}_k, \\ &= \mathbf{m}_k^T \,\mathrm{skew}(\mathbf{v})\,\mathbf{m}_k, \\ &= \mathbf{m}_k^T (\mathbf{v} \times \mathbf{m}_k), \\ &= \mathbf{0}. \end{aligned}$$

Clearly, displacements of the form

$$\mathbf{N} + \epsilon \mathbf{H}_t = \mathbf{N} + \epsilon \mathbf{v} \mathbf{1}_n^T$$

are pure translations and of the form

$$\mathbf{N} + \epsilon \mathbf{H}_r = \mathbf{N} + \epsilon \,\mathrm{skew}(\mathbf{v})\mathbf{N}$$

are pure infinitesimal rotations, since $\mathbf{v} \times \mathbf{n}_i$ is orthogonal to $\mathbf{n}_i$ for all i. Since all possible choices of $\mathbf{v} \in \mathbf{R}^3$ can be parametrized as linear combinations of three linearly independent vectors in $\mathbb{R}^3$, the above directions generally[2] produce six linearly independent directions $\mathbf{H}$ for which $\text{vec}(\mathbf{H})^T \mathbf{K}_\phi \, \text{vec}(\mathbf{H}) = 0$.

Notice also that

$$\mathbf{K}_\sigma \, \text{vec}(\mathbf{H}) = \text{vec}\left(\mathbf{H}\,\mathbf{C}^T \mathbf{\Sigma}(\mathbf{M})\mathbf{C}\right).$$

In particular, for $\mathbf{H} = \mathbf{H}_t$,

$$\mathbf{K}_\sigma \, \text{vec}(\mathbf{H}_t) = \text{vec}\left(\mathbf{v}\mathbf{1}_n^T \mathbf{C}^T \mathbf{\Sigma}(\mathbf{M})\mathbf{C}\right) = \mathbf{0},$$

since $\mathbf{1}_n^T \mathbf{C}^T = 0$. Furthermore, for $\mathbf{H} = \mathbf{H}_r$,

$$\mathbf{K}_\sigma \, \text{vec}(\mathbf{H}_r) = \text{vec}\left(\text{skew}(\mathbf{v})\mathbf{N}\mathbf{C}^T \mathbf{\Sigma}(\mathbf{M})\mathbf{C}\right) = \mathbf{0},$$

because $\mathbf{N}$ is assumed to be an equilibrium point, in which case we have $\mathbf{N}\mathbf{C}^T \mathbf{\Sigma}(\mathbf{M})\mathbf{C} = \mathbf{0}$.

The conclusion is that $\mathbf{K}\mathbf{h} = \mathbf{0}$ for any direction $\mathbf{h}$ spanned by vectors $\mathbf{H}_t$ and $\mathbf{H}_r$. In other words, $\text{vec}(\mathbf{H}_t)$ and $\text{vec}(\mathbf{H}_r)$ are eigenvectors of the matrix $\mathbf{K}$ associated with a null eigenvalue.

[2]The only exception being on the extremely particular case when $\mathbf{N} = \mathbf{v}\mathbf{1}_n^T$, which can be easily ruled out by inspection of $\mathbf{N}$.

Chapter 3

Design of Compressive Structures

In this chapter we determine the best *tensegrity configuration* and string *connectivity* to minimize mass of a structure in compression, subject to a constraint on buckling strength and subject to the use of self-similar iterations (where we replace a bar with yet another tensegrity system). Of course, using self-similar iterations to fill space might not yield minimal mass over all other possible arrangements of members that are not restricted by a self-similar rule, but the simplicity of the math and the beauty of the symmetry are attractive as a starting point for further optimizations of structures.

Self-similar concepts describe a repetitive process where a geometrical object is replaced by yet another similar geometrical object. *Fractals* are the results of self-similar iterations as the number of self-similar iterations approaches infinity. Fractals have a mathematical description, but are art forms as well, having intrigued both artists (M.C. Escher and Snelson) and scientists (Mandelson and Wolfram). Mathematicians have produced a rich theory to fill space with simple self-similar rules, under the labels of *fractals, tiling,* and *tessellations.* Part of the intrigue is the *simplicity* of the fundamental element that generates the fractal by a self-similar rule. Artists have generated impressive forms with straight lines (Snelson), or simple curves, repeated within imaginative rules. Scientists discover rules for filling space with repetitive simple geometries. One can wonder (with Stephen Wolfram) whether any of the repetitive rules generated by these man-made rules can match the geometries generated by nature (such as seashell material). In biological systems, one might argue that the geometry patterns in the material play a vital role in the generation of a mechanical, acoustical, or electrical property that enhances the survival of a species.

The literature on fractals is only about geometry, and no specific mechanical or electrical properties are associated with a material that might be

R.E. Skelton, M.C. de Oliveira, *Tensegrity Systems,*
DOI 10.1007/978-0-387-74242-7_3,

constructed according to this geometry. Our focus is the opposite. We shall study rules of filling space [Ben95, BS03] with a self-similar rule, while guaranteeing a specific mechanical property at each self-similar iteration. That is the goal of this chapter, in the special case where external forces place the structure in compression.

One can then study the specific mechanical properties underlying nature's biological choices of geometry. This represents a popular focus in bioengineering, where great attention is given to the mechanical response of cells and its components Cite [VSBS05]. Hence, the ironic union of interests of artists and engineers in space-filling geometry is spawned from opposite objectives, beauty of form versus function. Indeed, the requirements of stability, mechanical stiffness, and strength rule out *most* of the self-similar forms that can fill the space. The study of composite materials has demonstrated that severe limitations are required on the choice of internal geometrical forms to yield desired mechanical properties for the bulk material. The tensegrity paradigm for this geometry will serve our purpose both in form and function objectives.

To begin our study in finding a self-similar rule to fill space with specific mechanical properties, tensegrity structures are employed as the building elements, and we seek to fill a constrained space rather than the whole of space. Hence our fractals will fill a finite-dimensional space with self-similar structures of infinite complexity.

Tensile members stabilize the shape of a tensegrity structure; the tendons can be prestressed to provide robustness against uncertainties in external loads. Tensile members are more efficient in mass than compressive members. For this reason, tensegrity structures can be lightweight. Therefore, if the use of long compressive members is minimized, while the use of tensile members is maximized, one may be able to improve the strength and reduce the weight of the structure. In this case, one seeks to design a self-similar tensegrity structure such that it has the same strength but less mass than the original tensegrity structure.

In the next two chapters we shall answer two optimization questions: (1) *What is the minimal mass structure in compression?* (2) *What is the minimal mass structure in bending?* A tensegrity answer to these two questions will naturally appear, where the complexity of the final answer depends upon stiffness constraints, if required.

The replacement of the compressive members by another tensegrity structure yields a new structure with more compressive members (but shorter ones). If the replacement is continuously applied to the new structure produced by the previous replacement, this procedure results in a self-similar structure and the name *self-similar tensegrity* [CS02]. This self-similar process can be repeated indefinitely, and one would like to guarantee certain mechanical properties of the result. If the self-similar iterations continue ad infinitum, we call the result *tensegrity fractals.* With the self-similar idea, the structural members can be scaled down for easier manufacturing and construction. At the microscopic level, this becomes material design, rather than structure design. Hence, at the mathematical level, we do not differentiate between material design and structure design.

Another motivation for this class of structures is controllability, since the lengths of strings can easily be controlled using actuators or controllable materials like piezoelectric and shape memory alloy. The concept enables the design of large deployable space tensegrity structures that have a small stowed volume [Fur92], or the design of large movable civil buildings that can regulate the solar energy and the airflow that is received internal to the building. Indeed, one motivation for this book is to provide the analytical tools with which one can simultaneously minimize the mass of a structure and the energy required to control it to specified performance, but this task is beyond the scope of this chapter. See some early attempts to control tensegrity structures in [CDGP, Fur92, Han92b, Ped98, Sul99].

3.1 Self-Similar Structures in Compression

Consider the class 1 tensegrity structure in Figure 3.1. Note in this figure that by choosing different tendon lengths of the class 1 unit we call $C4T16$, one can produce three different configurations of interest, called the *Box* unit in Figure 3.2, the *T-Bar* unit, in Figure 3.3, and the *D-Bar* unit in Figure 3.4.

1. *Box*: The rectangular shape called the *Box* unit of dimension $h \times w$ results by choosing the corners of the box to be nodes at which the forces are applied. In this case two parallel bars have become the same bar, reducing the total bars from four to two, forming a class 1 tensegrity structure, as in Figure 3.2.

2. *T-Bar*: The configuration we call a *T-Bar* unit in Figure 3.3 results when we rotate the *Box* unit so that the external forces are applied only at one node on each end, generating a class 4 tensegrity structure by placing a ball joint at the intersection of the bars.

3. *D-Bar*: The configuration that results when we choose $h = w = 0$ will be called a *Diamond-Bar* unit, or more simply labeled a *D-Bar* unit (Figure 3.4). This is a class 2 tensegrity structure, but there is a string-to-string connection (joint) where the vertical and horizontal strings intersect.

Primals and Duals: By the words tensegrity *primal* and tensegrity *dual*, we refer to two tensegrity systems for which the bars/tendons of one system are replaced by tendons/bars to obtain the second system, as defined in Section 2.4.2. Hence, in Figure 3.3, if the *T-Bar* system is labeled the *primal* system, then the *D-Bar* system in Figure 3.4 is its *dual*.

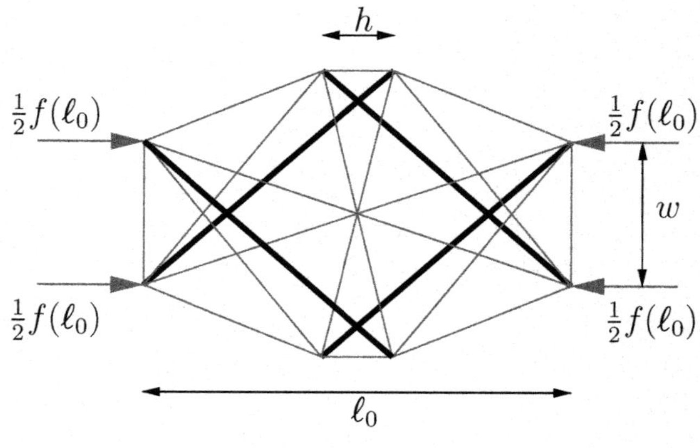

Figure 3.1: A $C4T16$ structure under critical compressive load, $f(\ell_0)$

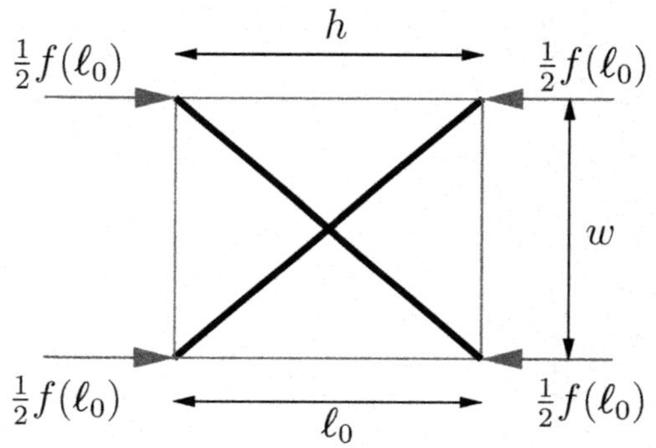

Figure 3.2: A *Box* structure under critical compressive load, $f(\ell_0)$

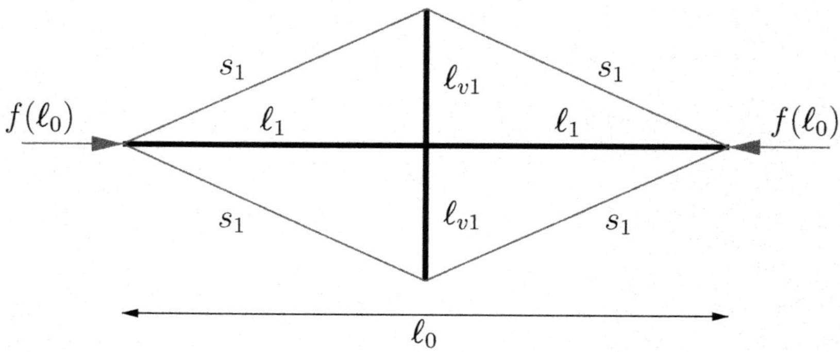

Figure 3.3: A *T-Bar* structure under critical compressive load, $f(\ell_0)$

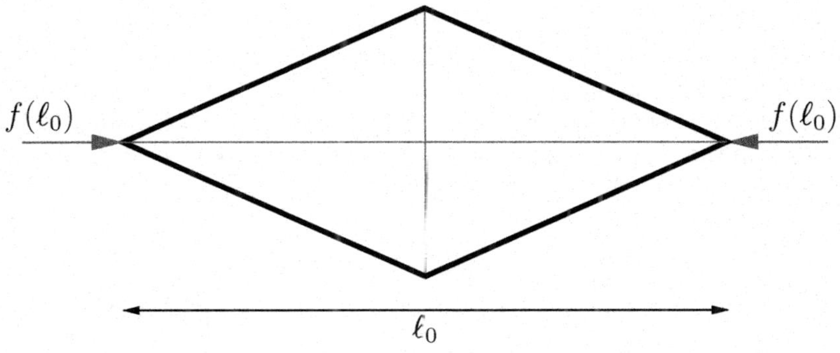

Figure 3.4: A *D-Bar* structure under critical compressive load, $f(\ell_0)$

3.1.1 Failure by Material Yielding

Structures may fail in two ways. The material may fail (material yields), or the structure may buckle (material bends). One may minimize the mass of a structure, subject to either buckling or yield constraints. We consider both modes of failure, but in this section we consider only yielding.

The mass $m(s)$ and yield strength $t(s)$ of a string of length s are related by

$$t(s) = \sigma A = \sigma \pi r^2, \quad m(s) = \rho \pi r^2 s, \tag{3.1}$$

where σ is the yield stress of the tendon material; r, s, $A = \pi r^2$ are, respectively, the radius, length, cross-sectional area, measured at yielding. From (3.1), it follows that

$$m(s) = c_s s t(s), \quad c_s = \frac{\rho_s}{\sigma_s}. \tag{3.2}$$

Likewise, yielding of compressive members of length ℓ, subject to force $f(\ell)$, has mass $m(\ell)$ given by $m(\ell) = c_b \ell f(\ell), \quad c_b = \rho_b/\sigma_b$. Now suppose bar and string material is the same. Then $c_s = c_b$ and the total system mass is proportional to J,

$$J = \sum_{i=1}^{m} \ell_i f(\ell_i), \tag{3.3}$$

where ℓ_i is the length of the ith member and $f(\ell_i)$ is the force in the ith member (in tension or compression). Indeed, in Section 2.6.1 we have already shown that when the same material is used for strings and bars the total mass of a structure is minimized under yielding constraints if a certain normalized material volume J which is independent of material properties (see (2.10)) is minimized. Under these circumstances, the optimal geometric arrangement of material (the topology) is independent of the choice of material.

We will now compute the value of J for the two cases: the primal *T-Bar* unit in Figure 3.3 and the dual *D-Bar* unit in Figure 3.4. It is convenient to label strings and bar members differently. We denote the length of bars by ℓ_i, the length of strings by s_i, the force in the bar of length ℓ_i by $f(\ell_i)$, and the force in the string of length s_i by $t(s_i)$. In the loaded case (external force $f(\ell_0)$ applied), the equilibrium conditions for the primal system (the *T-Bar*) yield

$$\begin{aligned} \ell_1 &= s_1 \cos\alpha_1 = \ell_0/2, & \ell_{v1} &= s_1 \sin\alpha_1, \\ f(\ell_{v1}) &= 2\sin\alpha_1 t(s_1), & f(\ell_1) &= f(\ell_0) + 2\cos\alpha_1 t(s_1). \end{aligned} \tag{3.4}$$

Normalize J by $J_0 = \ell_0 f(\ell_0)$ to get $\hat{J}$ defined by $\hat{J} = J/J_0$. This leads to

$$\begin{aligned} \hat{J} &= \frac{1}{\ell_0 f(\ell_0)}(2\ell_1 f(\ell_1) + 2\ell_{v1} f(\ell_{v1}) + 4s_1 t(s_1)) \\ &= 1 + 2\cos\alpha_1 \hat{t}(s_1) + \frac{2\sin^2\alpha_1 \hat{t}(s_1)}{\cos\alpha_1} + \frac{2\hat{t}(s_1)}{\cos\alpha_1} \\ &= 1 + \frac{4\hat{t}(s_1)}{\cos\alpha_1}, \end{aligned}$$

where $\hat{t}(s_1) = t(s_1)/f(\ell_0)$.
Note that, imposing yielding constraints only, the *T-Bar* unit cannot have mass less than the original bar, since $\hat{J} \geq 1$. So, if yielding were the method of failure, then a single bar is better than the *T-Bar* unit.

Note that since the *D-Bar* unit is the dual of the *T-Bar* unit, the equilibrium conditions (3.4) apply with a change in sign of the forces and replacing $(f(\ell_1), f(\ell_{v1}), t(s_1))$, respectively, by $(-t(s_1), -t(s_{v1}), -f(\ell_1))$. This leads to

$$\begin{aligned} s_1 &= \ell_1 \cos\alpha_1 = \ell_0/2, & s_{v1} &= \ell_1 \sin\alpha_1, \\ t(s_{v1}) &= 2\sin\alpha_1 f(\ell_1), & 2\cos\alpha_1 f(\ell_1) &= f(\ell_0) + t(s_1) \end{aligned} \tag{3.5}$$

and

$$\begin{aligned} \hat{J} &= \frac{1}{\ell_0 f(\ell_0)}(4\ell_1 f(\ell_1) + 2s_{v1} t(s_{v1}) + 2s_1 t(s_1)) \\ &= \frac{1}{\cos^2\alpha_1}(1 + \hat{t}(s_1)) + \hat{t}(s_1) + \frac{\sin^2\alpha_1}{\cos^2\alpha_1}(1 + \hat{t}(s_1)) \\ &= 1 + 2\tan^2\alpha_1 + \frac{2}{\cos^2\alpha_1}\hat{t}(s_1). \end{aligned}$$

Again, $\hat{J} \geq 1$ yielding the conclusion that, if yielding is the mode of failure under compressive loads, the single bar requires less mass than the *D-Bar* unit.

3.1.2 Buckling Constraints

When all members (bars and strings) are designed subject to yield constraints, the normalization above allows material properties to be eliminated from the minimal mass calculations. In this section we design strings subject to yield constraints, while bars will be subject to buckling constraints. Hence, the same normalization will not eliminate material properties from the total mass calculation.

Let a solid rod have length ℓ_0, radius r_0, mass density ρ_b, Young's modulus of elasticity E, and mass $m(\ell_0)$. According to Euler, a bar of length ℓ_0 under a compressive load $f(\ell_0)$ buckles if,

$$f(\ell_0) = \frac{E\pi^3 r_0^4}{4\ell_0^2}, \quad m(\ell_0) = \rho_b \pi r_0^2 \ell_0. \tag{3.6}$$

From (3.6) it follows that

$$m(\ell_0) = c_b \ell_0^2 \sqrt{f(\ell_0)}, \quad c_b = \frac{2\rho_b}{\sqrt{\pi E}}. \tag{3.7}$$

We pause for one calculation for a hollow cylinder, to make a point. For a hollow tube with inner radius r_i and outer radius r_0 the above equations become

$$f(\ell_0) = \frac{E\pi^3 (r_0^4 - r_i^4)}{4\ell_0^2}, \quad m(\ell_0) = \rho_b \pi (r_0^2 - r_i^2)\ell_0. \tag{3.8}$$

From (3.8) the mass of the hollow cylinder $m(\ell_0)$ satisfies a quadratic equation,

$$m^2(\ell_0) + 2\pi\rho_b \ell_0 r_i^2 m(\ell_0) - \frac{1}{\pi E} 4\rho_b^2 \ell_0^4 f(\ell_0) = 0, \tag{3.9}$$

which yields $m(\ell_0)$ approaching zero as the radius of the cylinder r_i approaches infinity. This fact suggests that, in the absence of material yielding or wall panel buckling, the buckling strength of a hollow cylinder approaches infinity (or the required mass of the hollow cylinder approaches zero), as the cylinder radius approaches infinity. Of course, panel buckling will occur in these cases, since the wall of the hollow cylinder becomes too thin. Hence, for practical reasons, a hollow cylinder is the minimal mass solution for compressive loads, if space is available to make the radius large enough. But in the presence of restricted space for the cylinder, we will show that the cylinder is not the minimal mass solution.

Mathematically we shall assume that the compressive members are *solid* cylinders, which, by a slight abuse of language, we call *bars*. Results very similar to those in this chapter can be obtained by using the more efficient *hollow* cylinders, but the ideas and the results are similar, and for our purposes the additional complexity of the math to treat hollow cylinders is unjustified here, so we shall present only the solid bar case.

An interesting fact to note, without explicit derivation, is that, under compressive loads $f(\ell_0)$, the *Box* unit in Figure 3.1 cannot have less mass than the original single bar, given that the same buckling strength is preserved. However, the addition of a *second Box* unit, overlapping with the first, creates the $C4T16$ unit in Figure 3.1 which *can* indeed reduce mass over the original bar. All the remaining three units, $C4T16$, *D-Bar*, and the *T-Bar* units in Figure 3.1, can be obtained as special cases of the $C4T16$ unit, by choosing appropriate string lengths in the $C4T16$ unit. Furthermore, each of these three units can save mass compared to the single bar unit, by appropriate choice of the free parameters describing the topology of each unit. One can therefore use any one of these three configurations as a self-similar rule to construct an efficient structure for taking compressive loads. Each self-similar iteration will reduce the combined mass of all the bars, without compromising strength (buckling load) of the structure. Shape constraints or fabrication issues might dictate which self-similar rule is better for a given application.

In this chapter, we will focus on the *T-Bar* rule and the *D-Bar* rule to develop self-similar structures that are efficient in compression. The $C4T16$ structure is less efficient than either the *D-Bar* unit (a class 2 tensegrity) or the *T-Bar* unit (a class 4 tensegrity), although $C4T16$ is a class 1 tensegrity that may have fabrication advantages. Below, these structures will be designed to match the compressive strength of the original single bar, while reducing the total bar mass.

The *dual* of the *T-Bar* system is the *D-Bar* system in Figure 3.1. We will show that for a suitable range of angle α both the primal and the dual can save mass over the original bar.

3.2 *T-Bar* Systems

3.2.1 The *T-Bar* Unit

Let a given solid cylinder (bar) of mass $m(\ell_0)$ buckle at load $f(\ell_0)$. Now replace the original cylinder by a *T-Bar* unit, where all eight members of the *T-Bar* unit are designed to fail at exactly the same value of the external load $f(\ell_0)$ as for the original bar of length ℓ_0. Hence, we say that the *strength* of the two systems is preserved. There are four strings in the *T-Bar* unit, each having length s_i. There are also four bars, two having length ℓ_1 and two having length ℓ_{v1}. We use the same name for the string label and the string length. That is, we say "string s_i has length s_i". Likewise for bars, we say "bar ℓ_i has length ℓ_i". We denote the tension in a string by using its label as the argument, so the tension in string s_1 is $t(s_1)$. Likewise, the force in a bar labeled ℓ_1 is $f(\ell_1)$, and the force in a bar with label ℓ_{v1} is $f(\ell_{v1})$. The tension $t(s_1)$ should be equal in all four tendons of our *T-Bar* unit. We will normalize $t(s_1)$ by the external force $f(\ell_0)$, and define $\hat{t}(s_1) = t(s_1)/f(\ell_0)$. We have two free parameters in this *T-Bar* design, the angle α_1 and the normalized string tension $\hat{t}(s_1)$. Let the mass of a bar of length ℓ_i be denoted by $m(\ell_i)$, where from (3.7) $m(\ell_i) = c_b\ell_i^2\sqrt{f(\ell_i)}$ and c_b is a material-dependent constant (for

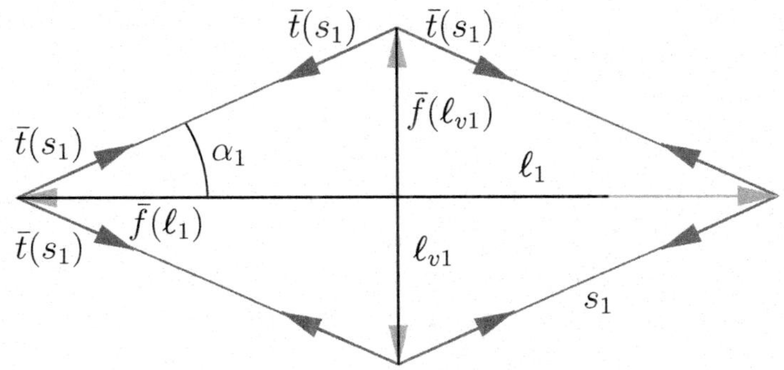

Figure 3.5: *T-Bar* forces for unloaded (prestressed) case, $f(\ell_0) = 0$

simplicity we assume the same material c_b for all bars, and the same material c_s for all strings).

We will require the same geometry for both the *unloaded* and *loaded* cases (hence the same ℓ_1, ℓ_{v1}, s_1). For string s_1 and bar ℓ_1 in Figure 3.3, the forces in the loaded case are denoted by $t(s_1)$ and $f(\ell_1)$. The forces in the unloaded case will be denoted by an overbar, namely, $\bar{t}(s_1)$ and $\bar{f}(\ell_1)$, as shown in Figure 3.5. Hence in the loaded case (3.4) applies, and in the unloaded case

$$\bar{f}(\ell_1) = f(\ell_0), \qquad \bar{f}(\ell_{v1}) = f(\ell_0)\tan\alpha_1, \qquad \bar{t}(s_1) = \frac{1}{2\cos\alpha_1} f(\ell_0).$$

In the loaded case (external force $f(\ell_0)$ applied), the forces in (3.4) apply, with $t(s_1) = 0$. These equations yield for the loaded case,

$$f(\ell_{v1}) = 0, \qquad f(\ell_1) = f(\ell_0), \qquad t(s_1) = 0.$$

Our goal is to obtain a stable system in both the loaded and unloaded structures, subject to the same configuration in both circumstances. In the loaded case the force $f(\ell_1)$ stabilizes the configuration, whereas in the unloaded case $\bar{t}(s_1)$ must stabilize. Hence, $\bar{t}(s_1)$ cannot be zero.

The mass assigned to a member is therefore the larger of the two masses, computed under both the loaded and unloaded cases. Hence, under buckling loads, the mass that must be assigned to the member of length ℓ_1 corresponds to the larger of the two forces $f(\ell_1)$ or $\bar{f}(\ell_1)$. That is, $m(\ell_1)$ is the larger of the two numbers $[c_b(\ell_1)^2(f(\ell_1))^{1/2}, c_b(\ell_1)^2(\bar{f}(\ell_1))^{1/2}]$. In the *T-Bar* unit, $f(\ell_1) = \bar{f}(\ell_1)$ and $f(\ell_{v1}) < \bar{f}(\ell_{v1})$. Hence, the mass of all bars in the *T-Bar* unit is m_{b_1}, where,

$$\begin{aligned} m_{b_1} &= 2m(\ell_{v1}) + 2m(\ell_1) \\ &= 2c_b(\ell_{v1})^2[(\bar{f}(\ell_{v1}))^{1/2} + (f(\ell_1))^{1/2}] \\ &= 2c_b(\ell_0/2)^2[\tan^2\alpha_1(f(\ell_0)\tan\alpha_1)^{1/2} + (f(\ell_0))^{1/2}] \\ &= 2c_b\,(\ell_0/2)^2\,(f(\ell_0))^{1/2}[\tan^{5/2}\alpha_1 + 1]. \end{aligned}$$

Hence, the mass of the *T-Bar* unit is less than the mass of the original bar if $\mu_{b_1} < 1$, where,

$$\mu_{b1} = \frac{m_{b_1}}{m(\ell_0)} = \frac{1}{2}[\tan^{5/2}\alpha_1 + 1],$$

and $\mu_{b_1} < 1$ if $\alpha_1 < 45°$.

Now let us add string mass. The string mass is $4m(s_1) = 4c_s s_1(max[t(s_1), \bar{t}(s_1)])$, where $t(s_1) = 0$ is the string tension in the externally loaded case and $\bar{t}(s_1) = f(\ell_0)/(2\cos\alpha_1)$ is the string tension in the externally unforced case. Therefore, the total mass ratio is

$$\begin{aligned}
\mu_1 &= \frac{1}{2}[\tan^{5/2}\alpha_1 + 1] + 4m(s_1)/m(\ell_0) \\
&= \frac{1}{2}[\tan^{5/2}\alpha_1 + 1] + 4\frac{c_s}{m(\ell_0)}\left(\frac{\ell_0}{2\cos\alpha_1}\right)\left(\frac{f(\ell_0)}{2\cos\alpha_1}\right) \\
&= \frac{1}{2}[\tan^{5/2}\alpha_1 + 1] + 4\left(\frac{c_s\sqrt{f(\ell_0)}}{c_b\ell_0}\right)\left(\frac{1}{2\cos\alpha_1}\right)^2 \\
&= \frac{1}{2}[\tan^{5/2}\alpha_1 + 1] + \epsilon(1 + \tan^2\alpha_1),
\end{aligned} \tag{3.10}$$

where

$$\epsilon = \frac{c_s\sqrt{f(\ell_0)}}{c_b\ell_0} = \frac{\rho_s\sqrt{\pi E f(\ell_0)}}{2\sigma_s\rho_b\ell_0} \tag{3.11}$$

is a dimensional parameter. We note that small ϵ corresponds to small external forces, soft bar material, high strength string, and long bars. Larger ϵ corresponds to large external forces, stiff bar material, low yield strength string, and short bars. In Figure 3.6, a plot of μ_1 versus α_1 is illustrated for various values of ϵ. Note that $\mu_1 \geq 1/2 + \epsilon$ so that for a mass reduction one should have $\epsilon \leq 1/2$, which yields a property between external force and string and bar material, specifically,

$$\frac{f(\ell_0)}{\ell_0^2} < \left(\frac{\rho_b^2}{\pi E}\right)\left(\frac{\sigma_s}{\rho_s}\right)^2. \tag{3.12}$$

Therefore, in any *T-Bar* self-similar rule for constructing compressive structures, the property of bars that is important for mass reduction is $\rho_b^2/(\pi E)$, and the property of strings that is important for mass reduction is σ_s/ρ_s.

In our numerical examples, we assume that both compressive members and tensile members are made of steel, with density $\rho_b = \rho_s = 7862\,\text{kg/m}^3$, Young's modulus $E = 2.06\times10^{11}$ N/m^2, and the yield strength $\sigma_s = 6.9\times10^8$ N/m^2. For example, (3.12) suggests that using the same steel for strings and bars, mass cannot be reduced by *T-Bar* self-similar iterations if $f(\ell_0)/\ell_0^2 > 0.001\,\text{N/m}^2$.

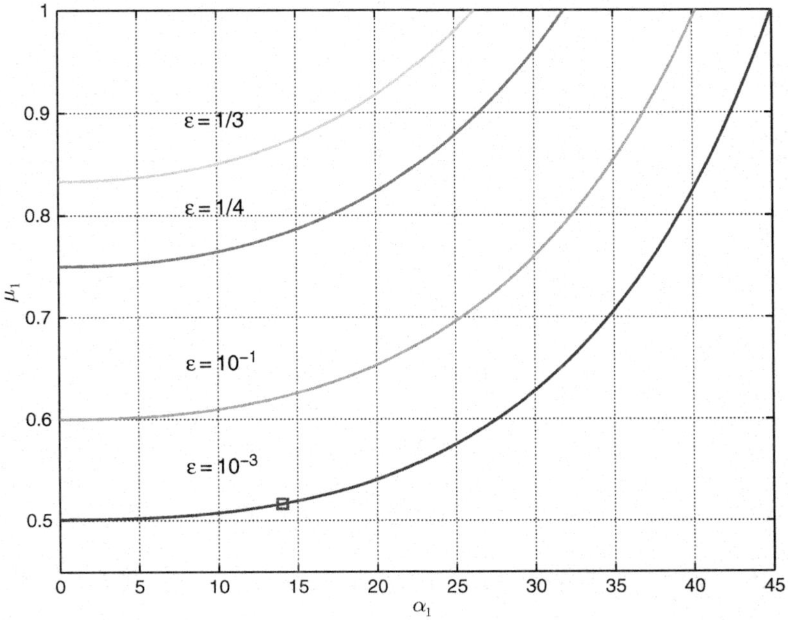

Figure 3.6: *T-Bar* unit: μ_1 versus α_1, from (3.10)

Example 3.1

For steel strings and bars $c_s/c_b = 0.5829 \times 10^{-3}$. With external force $f(\ell_0) = 2.942\,\ell_0^2$ N, we have $\epsilon = 0.001$ and from the above equation, choosing $\tan\alpha_1 = 0.25$, we obtain $\mu_1 = 0.517$. This point is marked with a square in Figure 3.6.

Example 3.2

Suppose a force f_1 is applied vertically downward at the top center node of Figure 3.5, in addition to the horizontal force f_0 we have already considered. In the presence of both horizontal loads f_0 and vertical loads f_1, the force in the top two strings will be zero, by our previous assumption that these two forces are zero in the absence of f_1. Using the same techniques as above show that when the *T-Bar* is loaded in both directions with forces f_0 and f_1, where $\zeta = f_1/f_0$, the mass ratio is given by

$$\mu_1(f_0, f_1) = \frac{1}{2}[\sqrt{1 + \zeta/2\tan\alpha} + \tan^2\alpha\sqrt{\zeta} + \epsilon(1 + \tan^2)(1 + 2\zeta)]. \quad (3.13)$$

It is obvious that minimal mass of the nominal design (3.10) occurs at $\alpha = 0$. However, this design cannot tolerate any side forces f_1 without buckling the system, even though no individual bar buckles. We imagine that this side force is a small fraction of the nominal force f_0 applied in the

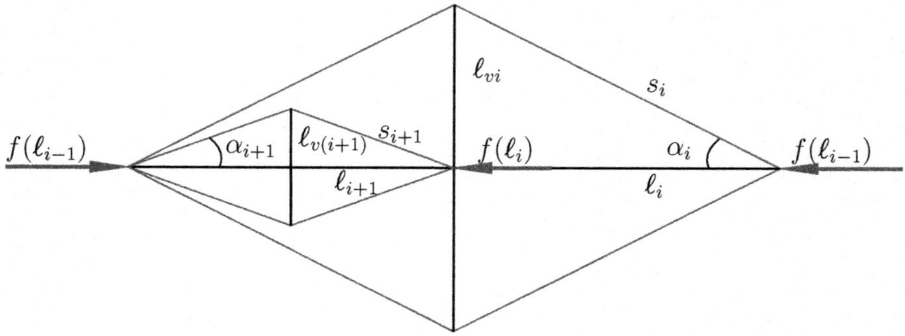

Figure 3.7: *T-Bar* self-similar iteration

horizontal directions. Hence $\zeta < 1$ and f_1 represents external disturbances ignored in the nominal design, but one can choose α to guarantee a certain degree of robustness to the external load environment.

3.2.2 The *T-Bar* Self-Similar Rule

We begin with a bar of length ℓ_0 compressed with the buckling load $f(\ell_0)$. Now we replace this bar by a *T-Bar* tensegrity unit of the same overall length, with the same external force $f(\ell_0)$ applied. Then we continue this process, replacing each bar with another *T-Bar* unit, designing all members of each unit to fail at the load applied to that unit. Repeating this self-similar process n times will preserve the same strength as the original bar. Now we evaluate the mass of the system after n iterations. The freedoms in each iteration are α_i and $\hat{t}(s_i)$ (see Figure 3.7). However, we will choose $\hat{t}(s_i)$ in the following way. When the external force $f(\ell_0)$ is applied, we will choose $\hat{t}_i = 0$, but when the external load is absent, we will choose a prestress that maintains the same load in bar ℓ_1 as in the loaded case. This means choosing prestress (the tension in strings s_1) to be $\bar{t}(s_1) = f(\ell_0)/(2\cos\alpha_1)$. For mass calculations, this requires that bars ℓ_i have the mass associated with the externally loaded case, and bars ℓ_{vi} and strings s_i will have mass dictated by the forces of the unloaded case (prestress). From Figure 3.7 note that

$$\ell_{i+1} = \frac{1}{2}\ell_i, \qquad s_i = \ell_i/\cos\alpha_i, \qquad \ell_{vi} = \ell_i \tan\alpha_i,$$
$$f(\ell_i) = f(\ell_{i-1}) + 2t(s_i)\cos\alpha_i, \quad f(\ell_{vi}) = 2t(s_i)\sin\alpha_i,$$

and

$$\frac{m(\ell_i)}{m(\ell_0)} = \left(\frac{\ell_i}{\ell_0}\right)^2 \sqrt{\frac{f(\ell_i)}{f(\ell_0)}}.$$

In case we choose $t(s_1) = 0$ in the above equations, we have the $f(\ell_i) = f(\ell_{i-1})$ and $f(\ell_{vi}) = 0$.

After n iterations the number of bars of length ℓ_n is 2^n, and the number of bars of lengths $\ell_{v1}, \ell_{v2}, \dots, \ell_{vn}$ sum to $\sum_{i=1}^{n} 2^i$ bars. Hence after n iterations, the total mass of the bars ℓ_n is given by

$$2^n m(\ell_n) = c_b 2^n (\ell_n)^2 \sqrt{f(\ell_n)} = c_b 2^n \left(\frac{\ell_0}{2^n}\right)^2 \sqrt{f(\ell_0)}.$$

Hence,

$$\frac{2^n m(\ell_n)}{m(\ell_0)} = \frac{1}{2^n}. \tag{3.14}$$

Now we must compute the mass of bars ℓ_{vi}, subject to force $\bar{f}(\ell_{vi})$, where $\bar{t}(s_i) = f(\ell_0)/(2\cos\alpha_i)$.

$$\begin{aligned} m(\ell_{vi}) &= c_b \ell_{vi}^2 \sqrt{f(\ell_{vi})} \\ &= c_b \left(\frac{1}{2^i}\ell_0 \tan\alpha_i\right)^2 \sqrt{f(\ell_0)\tan\alpha_i}. \end{aligned}$$

Hence,

$$\sum_{i=1}^{n} 2^i m(\ell_{vi})/m(\ell_0) = \sum_{i=1}^{n} 2^i \left(\frac{\tan^{5/2}\alpha_i}{2^{2i}}\right) = \sum_{i=1}^{n} \frac{\tan^{5/2}\alpha_i}{2^i}. \tag{3.15}$$

Now we must compute the mass of the strings s_i.

$$\begin{aligned} \sum_{i=1}^{n} 2^{i+1} m(s_i)/m(\ell_0) &= \sum_{i=1}^{n} 2^{i+1} s_i \bar{t}(s_i)/m(\ell_0) \\ &= \sum_{i=1}^{n} 2^{i+1} \frac{c_s}{m(\ell_0)} \left(\frac{\ell_0}{2^i \cos\alpha_i}\right)\left(\frac{f(\ell_0)}{2\cos\alpha_i}\right) \\ &= \sum_{i=1}^{n} 2^{i+1} \left(\frac{c_s\sqrt{f(\ell_0)}}{c_b \ell_0}\right)\left(\frac{1}{2^{i+1}\cos^2\alpha_i}\right) \\ &= \sum_{i=1}^{n} \epsilon(1+\tan^2\alpha_i), \end{aligned}$$

where ϵ is as in (3.11). Now the total mass after n self-similar iterations is the sum of the bar mass by adding (3.14) and (3.15). The total string mass is given by (3.14). Hence the total mass after n iterations is

$$\mu_n = m_n/m(\ell_0) = \frac{1}{2^n} + \sum_{i=1}^{n} \frac{\tan^{5/2}\alpha_i}{2^i} + \sum_{i=1}^{n} \epsilon(1+\tan^2\alpha_i). \tag{3.16}$$

In the special case where we choose the same $\alpha_i = \alpha$, for all i, then the total mass ratio is

$$\mu_n(\alpha_i = \alpha) = 2^{-n} + \tan^{5/2}\alpha(1-2^{-n}) + n\epsilon(1+\tan^2\alpha), \tag{3.17}$$

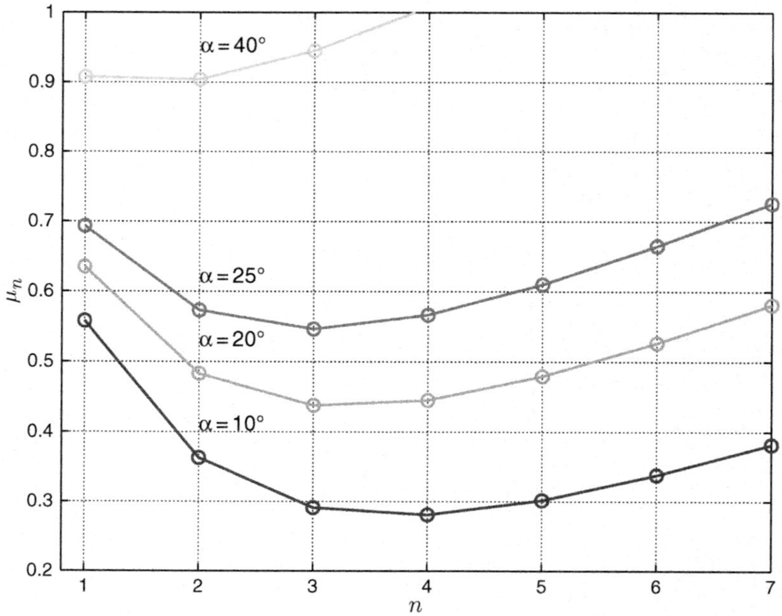

Figure 3.8: *T-Bar* unit: μ_n versus n, with $\epsilon = 0.05$, from (3.17)

where we have used the fact

$$\sum_{i=m}^{n} \beta^i = \frac{\beta^m - \beta^{n+1}}{(1-\beta)}, \quad \beta \neq 1 \tag{3.18}$$

to obtain the expression

$$\sum_{i=1}^{n} \frac{1}{2^i} = 1 - \frac{1}{2^n}, \tag{3.19}$$

which follows from the power series (3.18) by substituting $\beta = 1/2$ and $m = 1$. To reduce mass (3.17) requires $\mu_n \leq 1$, hence, $(2^{-n}-1)(1-\tan^{5/2}\alpha)+n\epsilon(1+\tan^2\alpha) \leq 0$, hence $\tan^{5/2}\alpha \leq 1$, hence $\alpha \leq 45°$.

A plot of $\mu_n(\alpha_i = \alpha)$ versus n appears in Figure 3.8. As n approaches infinity, tendon mass approaches infinity and the bar mass ratio approaches the value $(\tan^{5/2}\alpha)$. Clearly there exists a minimum mass at a finite number of iterations n.

To find the number of iterations that yield minimal system mass, differentiate (3.17) with respect to n and set this expression to zero. Solving this equation for n yields the optimal value n^*, satisfying,

$$2^{n^*} = \frac{\ln 2(1-\tan^{5/2}\alpha)}{(1+\tan^2\alpha)\epsilon}, \tag{3.20}$$

or, explicitly,

$$n^* = \lfloor \frac{\ln[(\ln 2)(1 - \tan^{5/2}\alpha)] - \ln[(1 + \tan^2\alpha)\epsilon]}{\ln 2} \rfloor, \tag{3.21}$$

where $\lfloor M \rfloor$ means round down the number M to the closest integer. Equation (3.21) yields the optimal *complexity* n^* of a structure in compression, when all $\alpha_i = \alpha$. The complexity is represented by n^*, since the number of tensegrity components required to build the minimal mass structure in compression is q, given by $q = 2^{n^*} + \sum_{i=1}^{n^*}(2^i + 2^{i+1}) = 3(2^{n^*}) - 2$.

Example 3.3

For steel strings and bars $c_s/c_b = 0.5829 \times 10^{-3}$. With external force $f(\ell_0) = 2.942\ell_0^2$ N, we have $\epsilon = 0.001$, and from the above equation, choosing $n = 6$ and $\tan\alpha = 0.25$, we obtain $\mu_6 = 0.0528$, indicating a structure of $q = 190$ components that has about 1/20th the mass of the original bar of length ℓ_0 and the same buckling strength.

Example 3.4

Verify, using (3.21), that the optimal mass in the above example occurs with a *T-Bar* structure of complexity $n^* = 9$, in which case the minimal mass is 0.0427 times the mass of the original bar, given by (3.17) using $n^* = 9$.

3.2.3 Optimal Column with Constant Width

A more lengthy example is developed here. Using the above self-similar rule with constant α yields the topology of Figure 3.9. Suppose we desire a compressive structure of length ℓ_0 to have uniform width w, as shown in Figure 3.10. The width constraint fixes the choice of α_i in (3.16). In our example we will choose α_i so that the ends of the ℓ_{vi} bars lie on the boundary of the shape constraint (constant width), until $i = k$, where k is the largest integer such that $\alpha_k \leq 45°$ (since mass is not reduced with larger α). For iterations $n \geq i > k$ mass cannot be reduced if the boundaries of the allowed shape (width) are reached by the bars ℓ_{vi}. Thus, the remaining α_i for $i > k$ will be chosen for mass reduction only and the bars ℓ_{vi} will not extend to the outside boundary of the structure. In fact for these remaining α_i, $i > k$, we will choose the most mass-efficient design among the set of α that have been used in the design. That is, $\alpha_i = \alpha_1 = \tan^{-1}(w/\ell_0)$, for $n \geq i > k$.

Now define z by $z := \sqrt{\ell_0/w}$. From the constant-width constraint,

$$\tan\alpha_i = 2^{i-1}w/\ell_0 = 2^{i-1}z^{-2}, \qquad 1 \leq i \leq \min\{k, n\},$$

where k is the largest i for which $\tan\alpha_i \leq 1$. Then, for any given z,

$$k = 1 + \lfloor \frac{\ln(\ell_0/w)}{\ln 2} \rfloor = 1 + \lfloor \frac{\ln(z^2)}{\ln 2} \rfloor. \tag{3.22}$$

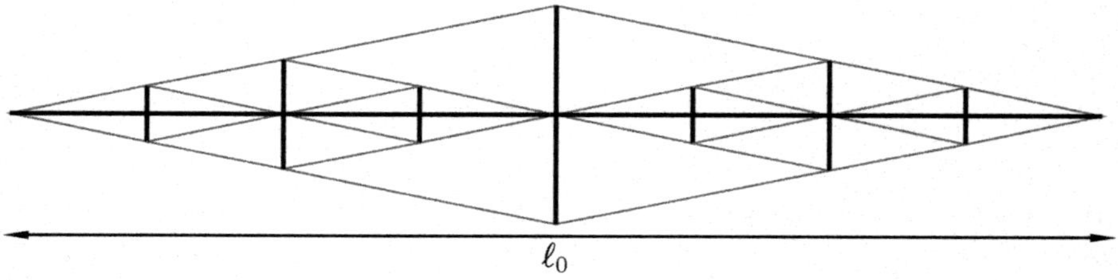

Figure 3.9: *T-Bar* self-similar $n = 4$, constant α

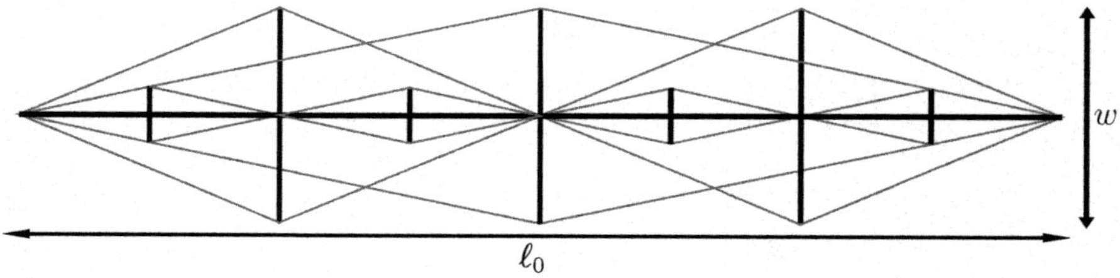

Figure 3.10: Constant-width *T-Bar* column $\ell_0/w = 2.3$, $n = 3$, $k = 2$, $[\alpha_i \leq 45^o,\ i = 1, \ldots, k]$, and $[\alpha_i = \alpha_1,\ i = k+1, \ldots, n]$

Using (3.16), (3.18), and $\tan\alpha_1 = z^{-2}$, we have

$$\mu_n = \begin{cases} \phi(n,k), & n > k; \\ \phi(n,n), & n \leq k; \end{cases} \tag{3.23}$$

where

$$\begin{aligned} \phi(n,k) &= 2^{-n} + \sum_{i=1}^{k} \left[2^{-i}\tan^{5/2}\alpha_i + \epsilon(1+\tan^2\alpha_i)\right] + \\ &\qquad \sum_{i=k+1}^{n} \left[2^{-i}\tan^{5/2}\alpha_1 + \epsilon(1+\tan^2\alpha_1)\right] \\ &= 2^{-n} + n\epsilon + \left(\frac{2^{3k/2}-1}{2(2^{3/2}-1)} + 2^{-k} - 2^{-n}\right) z^{-5} + \\ &\qquad \epsilon z^{-4}\left(\frac{4^k-1}{3} + n - k\right). \end{aligned}$$

Example 3.5

In Figure 3.11, the solid lines plot μ_n as in (3.23) for various values of ℓ_0/w and $\epsilon = 0.001$. The dashed lines are plots of $\phi(n,n)$, which represent the mass of a structure where the angles α_i are allowed to exceed $45°$ even when

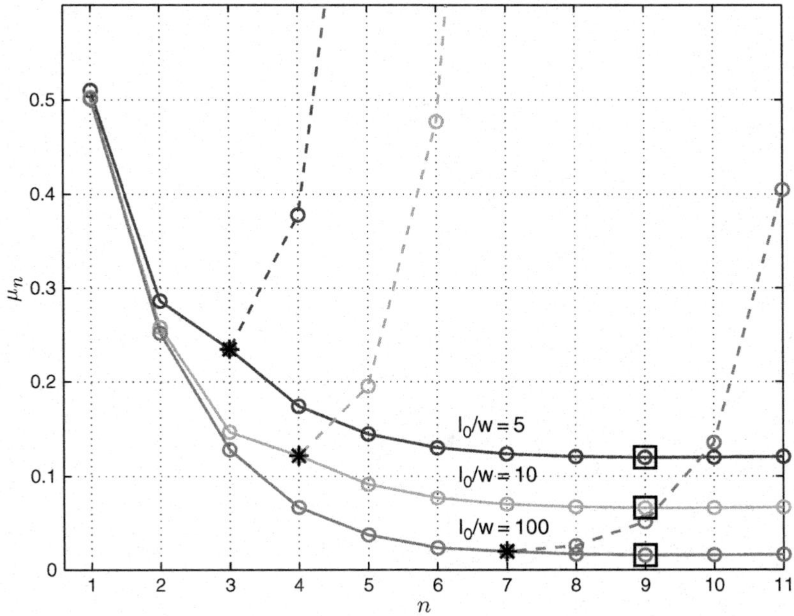

Figure 3.11: Constant-width *T-Bar* column: μ_n versus n, with $\epsilon = 0.001$, from (3.23)

$n \geq i > k$. Hence, for $n \leq k$ the two curves coincide splitting at the value marked by $*$ ($n = k$). The minimal mass design is indicated by squares which, in this example, are all at $n^* = 9$. For all practical purposes one can pick the less complex $n = 5$ instead of the optimal $n^* = 9$ with little impact on the mass.

Example 3.6

In Figure 3.12, the solid lines plot μ_n and $\phi(n, n)$ as in (3.23) for various values of ϵ and $\ell_0/w = 10$, as in Example 3.5. Because k does not depend on ϵ, $k = 4$ for all plots. Note that the minimal mass (squares) may be below, above, or equal to $n = k$, at which point the solid curve is not differentiable. Hence the minimum complexity n^* cannot generally be obtained by differentiation of (3.23). Indeed, the slope of the function μ_n is zero at $n = 3$ for the red curve ($\epsilon = 0.05$), which is indeed the global minimum. However, the slope is also zero at $n = 3$ for the green curve ($\epsilon = 0.02$), but this is a local and not the global minimum (which occurs at $n^* = 5$).

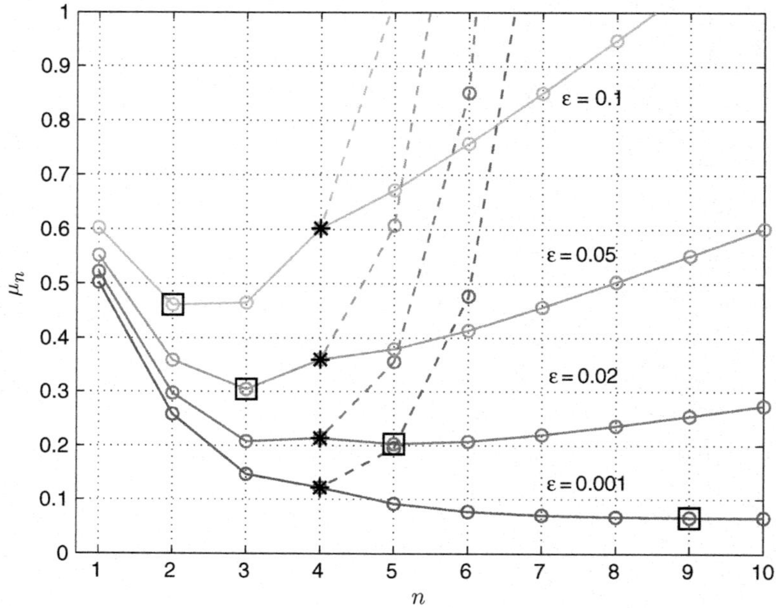

Figure 3.12: Constant-width *T-Bar* column: μ_n versus n, $\ell_0/w = 10$, from (3.23)

Example 3.7

If string mass is neglected, that is, $\epsilon = 0$, and the optimal number of iterations $n^* \to \infty$, then

$$\begin{aligned}\min_{n\geq 1} \phi(n,k) &= \left(\frac{2^{3k/2}-1}{2(2^{3/2}-1)} + 2^{-k}\right) + \lim_{n\to\infty}(1-z^{-5})2^{-n} \\ &= \left(\frac{2^{3k/2}-1}{2(2^{3/2}-1)} + 2^{-k}\right),\end{aligned}$$

because $z > 1$.

Example 3.8

For $z^2 = \ell_0/w = 10$ and $\epsilon = 0.02$ then, from Figure 3.12, $n^* = 5$ and $k = 4$. Figure 3.13 shows the optimal column.

The above results have not applied the *T-Bar* self-similar iterations to the ℓ_{vi} bars, but this can be done in a similar manner. These ℓ_{vi} bars become significant in mass only if the chosen tensions $t(s_i)$ are large, since only then can the forces $f(\ell_{vi}) = 2t(s_i)\sin\alpha_i$ become large. In examples we find that the mass of the ℓ_{vi} bars is insignificant for the range of values $\hat{t}(s_i) \leq 0.05$, $\alpha_i \leq \pi/4$.

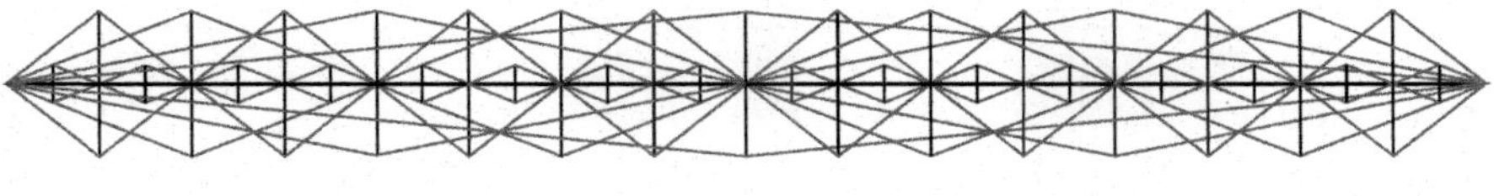

Figure 3.13: Optimal constant-width *T-Bar* column $\ell_0/w = 10$, $n^* = 5$, and $k = 4$

3.2.4 Yielding in *T-Bar* Self-Similar Systems

The bar ℓ_n yields when the bar force is $f(\ell_n) = \sigma_b \pi r_n^2$, or, equivalently

$$r_n^2 = f(\ell_n)/(\pi\sigma_b) = f(\ell_0)/(\pi\sigma_b). \tag{3.24}$$

The bar buckles when the bar force is $f(\ell_n) = f(\ell_0) = \pi^3 E r_n^4/(4\ell_n^2)$, or equivalently, using the fact $\ell_n = \ell_0/(2^n)$, we have

$$r_n^4 = 4\ell_n^2 f(\ell_n)/(\pi^3 E) = 4f(\ell_0)\ell_0^2/(2^{2n}\pi^3 E). \tag{3.25}$$

Equating (3.24) to the square root of (3.25) yields the iteration number n^{**} at which the buckling force and the yield force are the same on bar ℓ_n.

$$2^{n^{**}} = \frac{2\ell_0\sigma_b}{\sqrt{\pi E f(\ell_0)}} = \frac{(\rho_s/\sigma_s)}{(\rho_b/\sigma_b)}\frac{1}{\epsilon}. \tag{3.26}$$

For $n < n^{**}$, the failure of bar ℓ_n is by buckling, and (3.17) applies. If $n > n^{**}$ the failure of bar is by yielding, and then (3.17) does not apply.

It is straightforward to take the ln of (3.26) to find the explicit value of n^{**}. Figure 3.14 shows the range of iterations that are optimal for a constant angle *T-Bar* column, considering both yield and buckling type of failures when the same material is used for bars and strings.

For designs with constant angles α_i, as in Figure 3.9, the optimal complexity n is the smaller of the two numbers n computed from (3.20) and (3.26). For constant-width designs, as in Figure 3.13, the optimal complexity n is the smaller of the two numbers n computed from (3.23) and (3.26).

Example 3.9

For *T-Bar* columns with steel materials, the number of self-similar iterations before yielding satisfies $2^{n^{**}} = 232.9 f(\ell_0)^{-1/4}$, or equivalently, $n^{**} = [\ln 233 - (\ln f(\ell_0))/4]/\ln 2 = 7.86 - 0.36 \ln f(\ell_0)$, where $f(\ell_0)$ is the external force applied. Hence, for steel materials with $f(\ell_0) = 1$, the largest number of self-similar iterations before yielding is 7.

3.2.5 Three-Dimensional *T-Bar* System

Three-dimensional versions of the *T-Bar* unit can be built as shown in Figure 3.15. The structure in Figure 3.15 has $N = 3$ bars of the ℓ_v type.

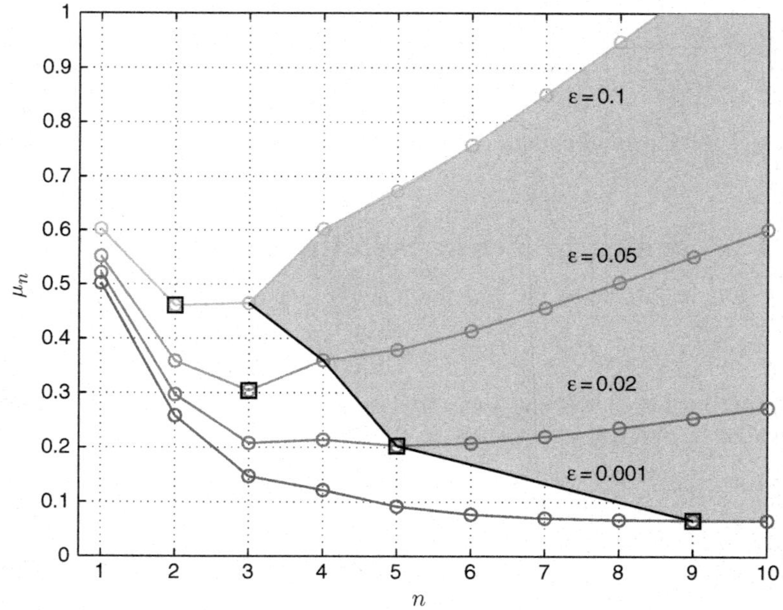

Figure 3.14: Constant-width *T-Bar* column: μ_n versus n, $\ell_0/w = 10$, from (3.23); green (shaded) area is where yielding is the mode of failure

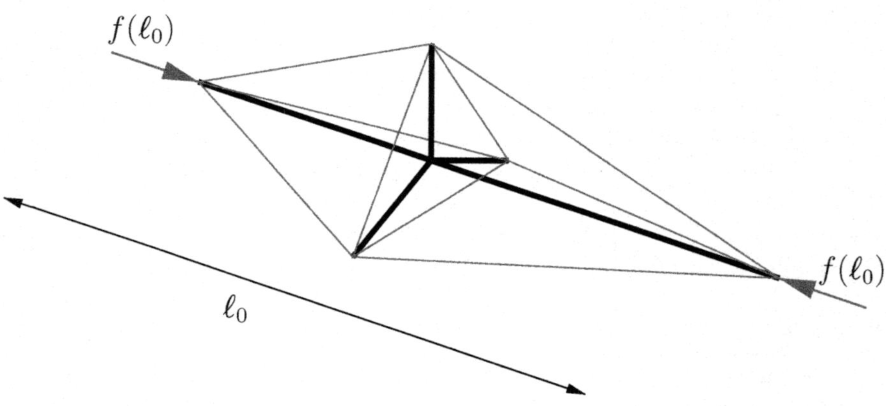

Figure 3.15: Three-dimensional *T-Bar* system

In general, one could build three-dimensional *T-Bar* units with any number N of bars of the ℓ_v type. The $N = 3$ type has a triangular cross-section and is the most efficient. The $N = 4$ unit has a square cross-section and might be preferred in certain envelope constraints. The same analysis as above will yield the self-similar realizations of optimal compressive structures in three dimensions. The math is the same as for the planar system, except there are

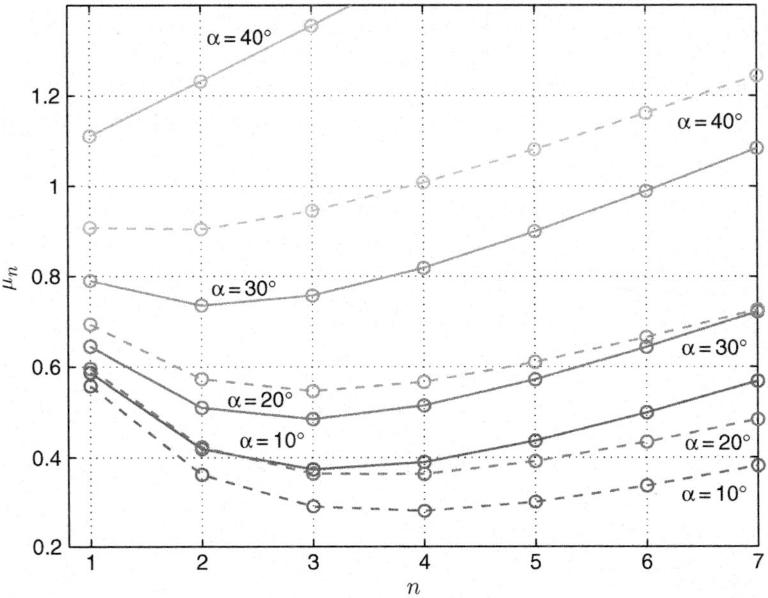

Figure 3.16: *T-Bar* unit: μ_n versus n, with $\epsilon = 0.05$ for planar ($N = 2$, *dashed lines*) and three-dimensional ($N = 3$, *solid lines*) cases, from (3.17) and (3.28)

N times as many strings s_i and three times as many bars ℓ_{vi}. Hence, (3.16) reveals the mass for the three-dimensional *T-Bar* with N bar of the ℓ_v-type self-similar compressive structure after n iterations to be

$$\mu_n^N = \frac{1}{2^n} + \frac{N}{2}\sum_{i=1}^{n} \frac{\tan^{5/2}\alpha_i}{2^i} + \frac{N}{2}\sum_{i=1}^{n} \epsilon(1 + \tan^2\alpha_i). \tag{3.27}$$

By following the same steps that followed (3.16) in the above section, one can determine the optimal number of iterations, n^*, for the three-dimensional *T-Bar* systems, with constant α or with constant diameter w.

For the three-dimensional case with $N = 3$ and constant α, the above equation reduces to

$$\mu_n^3(\alpha_i = \alpha) = 2^{-n} + (3/2)\tan^{5/2}\alpha(1 - 2^{-n}) + (3/2)n\epsilon(1 + \tan^2\alpha). \tag{3.28}$$

This three-dimensional case with $N = 3$ is shown in Figure 3.16 by solid lines. For comparison, the case $N = 2$, plotted in Figure 3.8, is shown by dashed lines.

Note that the price for three-dimensional stability, over the planar case, is less mass savings for a given α, or else α must be chosen at least 4.63° smaller. That is, for the planar case $\alpha < 45°$ was required to reduce mass, but in the three-dimensional case $\tan\alpha < (2/N)^{2/5}$. For $N = 3$ this means that $\alpha < 40.374°$. This follows from (3.27) by requiring $\mu_n^N \leq 1$ when $\epsilon = 0$, since in this case, $(2^{-n}-1)(1-(N/2)\tan^{5/2}\alpha) \leq 0$, requiring $\tan^{5/2}\alpha \leq 2/N$.

For any N, the minimal mass is achieved by finding the iteration number $n = n^*$ that minimizes

$$\mu_n^N(\alpha_i = \alpha) = 2^{-n} + (N/2)\tan^{5/2}\alpha(1 - 2^{-n}) + (N/2)n\epsilon(1 + \tan^2\alpha). \tag{3.29}$$

Differentiating this with respect to n and solving for n^* yields

$$2^{n^*} = \frac{2\ln 2(1 - \frac{N}{2}\tan^{5/2}\alpha)}{\epsilon N(1 + \tan^2\alpha)}. \tag{3.30}$$

Now substitute (3.30) into (3.29) to get the minimal mass ratio μ_{n^*}

$$\mu_{n^*}^N = \frac{N}{2}\tan^{5/2} + \epsilon\frac{N(1+\tan^2\alpha)}{2\ln 2}\left[1 + \ln\left(\frac{2\ln 2(1 - \frac{N}{2}\tan^5/2\alpha)}{\epsilon N(1+\tan^2\alpha)}\right)\right].$$

Note that the total bar mass of the optimal compressive structure is simply $m_b = (\frac{N}{2}\tan^{5/2}\alpha)m(\ell_0)$. Thus, for any chosen α, the smallest mass for a stable three-dimensional compressive structure is for $N = 3$.

3.3 *D-Bar* Systems

3.3.1 The *D-Bar* Unit

In this section we replace the original bar of mass $m(\ell_0)$ and length ℓ_0 by the *dual* of the *T-Bar* system, which is called the *D-Bar* system in Figure 3.4. We will replace the original bar with *D-Bar* units to get a structure with the same compressive strength.

For the *D-Bar* unit the length of all four bars is ℓ_1, and the lengths of the two strings are s_1 and s_{v1}. In general, for the ith iteration, with specified $f(\ell_0)$ and ℓ_0, the member lengths and forces are given by

$$\ell_i\cos\alpha_i = s_i, \qquad \ell_i\sin\alpha_i = s_{vi}, \tag{3.31}$$

$$2f(\ell_i)\cos\alpha_i = f(\ell_{i-1}) + t(s_i), \qquad 2f(\ell_i)\sin\alpha_i = t(s_{vi}), \tag{3.32}$$

where

$$\ell_1 = \ell_0/(2\cos\alpha_1).$$

When $i = 1$ we get the single *D-Bar* unit of Figure 3.4 and $i = 3, \ldots, 6$ in Figure 3.17. See also Figure 3.18 for the case $n = 2$ with $\alpha_1 > \alpha_2$.

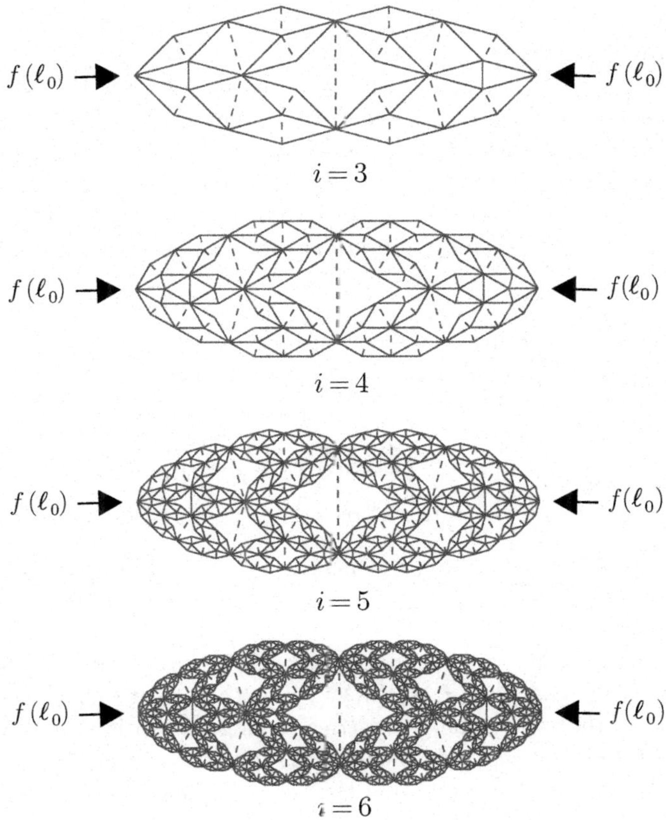

Figure 3.17: Configurations of the planar *D-Bar* self-similar structure with constant $\alpha = 15°$. The s_{vi} strings are in red (dashed lines). The s_i strings are not shown since they take no tension in these critical states

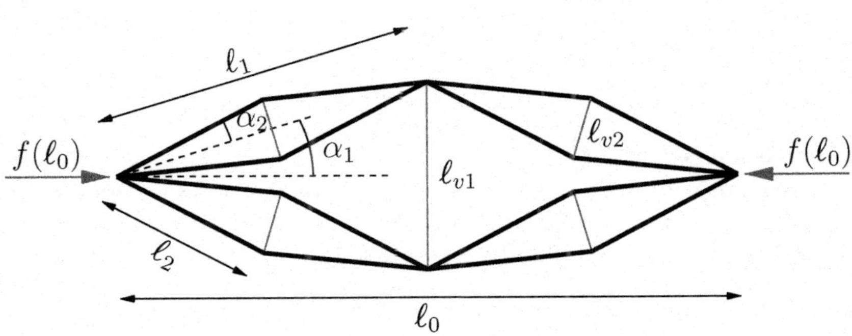

Figure 3.18: A *D-Bar*, $n = 2$ structure with $\alpha_1 > \alpha_2$

One is free to choose α_i and the tension $t(s_i)$. For smallest mass in the loaded case we choose $t(s_i) = 0$, but the mass of string s_i will not be zero, since we will use the tension in the unforced case to size this string. From previous discussions the string mass of s_i is based upon the largest tension in the string, between loaded and unloaded units. That is, for our choice $t(s_i) = 0$, we have $max[t(s_i), \bar{t}(s_i)] = \bar{t}(s_i)$. The total bar mass of the *D-Bar* unit is $m_{b_1} = 4m(\ell_1)$. The ratio of the bar mass of the *D-Bar* unit (composed of four bars of length ℓ_1 and mass $m(\ell_1)$) and the mass of the original bar of length ℓ_0 is

$$4\frac{m(\ell_1)}{m(\ell_0)} = 4\left(\frac{\ell_1}{\ell_0}\right)^2 \sqrt{\frac{f(\ell_1)}{f(\ell_0)}}, \tag{3.33}$$

$$= 4\left(\frac{1}{2\cos\alpha_1}\right)^2 \sqrt{\frac{f(\ell_0) + t(s_1)}{2\cos\alpha_1 f(\ell_0)}} \tag{3.34}$$

$$= (2\cos^5\alpha_1)^{-1/2}, \tag{3.35}$$

which has a lower bound of $1/\sqrt{2}$ when $t(s_1) = 0$ is chosen. Observe that this mass ratio cannot be less than 1 unless $\sqrt{2}(\cos^{5/2}\alpha_1) > 1$, or equivalently, $\alpha_1 < 29.477°$.

Now we must add string mass to the *D-Bar* unit. Recall the mass of the string is $m(s_1) = (\rho_s/\sigma_s)s_1 t(s_1) = c_s s_1 t(s_1)$. Note from the unloaded case in (3.31) that the same geometrical configuration is maintained in the unloaded and loaded cases if we choose the prestress to be $\bar{t}(s_1) = f(\ell_0)$. In the unloaded case we have $\bar{t}(s_{v1}) = \bar{t}(s_1)\tan\alpha_1$. Hence the masses of strings s_1 and s_{v1} are derived as follows:

$$m(s_1) = c_s s_1 \bar{t}(s_1) = c_s \ell_1 \cos\alpha_1 f(\ell_0) = c_s \ell_0 f(\ell_0)/2$$

and

$$m(s_{v1}) = c_s s_{v1} \bar{t}(s_{v1}) = c_s \ell_1 \sin\alpha_1 \bar{t}(s_1)\tan\alpha_1 = c_s \ell_0 f(\ell_0)\tan^2\alpha_1/2.$$

For the *D-Bar* system, the total string mass $m_{s_1} = 2m(s_1) + 2m(s_{v1})$. Therefore, the total string mass for the *D-Bar* unit is

$$2\left[m(s_1) + m(s_{v1})\right] = \epsilon(1 + \tan^2\alpha_1)m(\ell_0), \tag{3.36}$$

where ϵ is defined as in (3.11).

Herein we will always assume all strings are the same material, and all bars are the same material. The total mass of the *D-Bar* unit is therefore the sum of bar mass (3.35) and string mass (3.36),

$$m_1 = 4m(\ell_1) + 2m(s_1) + 2m(s_{v1}) = \mu_1 m(\ell_0),$$

hence

$$\mu_1 = (2\cos^5\alpha_1)^{-1/2} + \epsilon(1 + \tan^2\alpha_1). \tag{3.37}$$

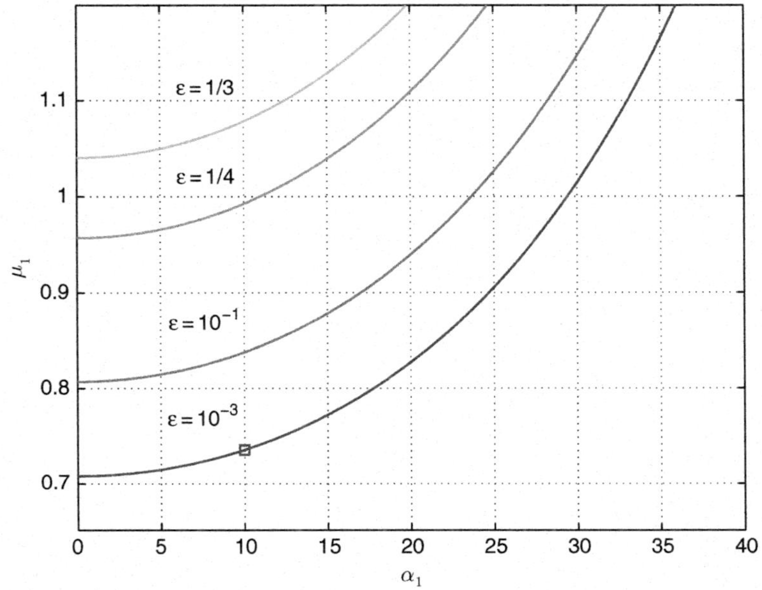

Figure 3.19: *D-Bar* unit: μ_1 versus α_1, from (3.37)

In Figure 3.19, a plot of μ_1 versus α_1 is illustrated for various values of ϵ. Note that $\mu_1 \geq \sqrt{2}/2 + \epsilon$ (compare with $\mu_1 \geq 1/2 + \epsilon$ in the case of a *T-Bar* unit) so that for mass reduction one should have $\epsilon \leq 1 - \sqrt{2}/2 \approx 0.3$. This is why no mass reduction is possible for $\epsilon = 1/3$ in Figure 3.19.

Example 3.10

For steel strings and bars and $f(\ell_0) = 2.942\ell_0^2$, as in previous examples, we have $\epsilon = 0.001$. Then for $\alpha_1 = 10°$, $\mu_1 = 0.73$ (compare with $\mu_1 = 0.517$ for the *T-Bar* design). This point is marked with a square in Figure 3.19.

3.3.2 The *D-Bar* Self-Similar Rule

Using (3.31) it is straightforward to show that the force in the bars of length ℓ_n at the nth iteration is

$$\frac{f(\ell_n)}{f(\ell_0)} = \frac{1}{\prod_{i=1}^{n} (2\cos\alpha_i)}. \tag{3.38}$$

Hence, after the nth iteration, the mass of a bar of length ℓ_n (normalized by the mass of bar of length ℓ_0) is

$$\frac{m(\ell_n)}{m(\ell_0)} = \left(\frac{\ell_n}{\ell_0}\right)^2 \sqrt{\frac{f(\ell_n)}{f(\ell_0)}} = \left(\frac{1}{\prod_{i=1}^{n}(2\cos\alpha_i)}\right)^{5/2}. \tag{3.39}$$

Using the fact that the loaded case yields $t(s_{v1}) = t(s_i) = 0$, but $\bar{t}(s_i) = \bar{t}(s_i)\tan\alpha_i$, and hence, after the nth iteration, the forces in strings s_n and s_{vn} are

$$\bar{t}(s_n) = f(\ell_{n-1}) = \frac{f(\ell_0)}{\prod_{i=1}^{n-1}(2\cos\alpha_i)},$$

$$\bar{t}(s_{vn}) = \bar{t}(s_n)\tan\alpha_n = 2\sin\alpha_n f(\ell_n) = \frac{f(\ell_0)\tan\alpha_n}{\prod_{i=1}^{n-1}(2\cos\alpha_i)}.$$

The masses of these strings are, at any iteration i,

$$\begin{aligned} 2m(s_i) + 2m(s_{vi}) &= 2c_s[s_i\bar{t}(s_i) + s_{vi}\bar{t}(s_{vi})] \\ &= c_s\ell_0 f(\ell_0)(1+\tan^2\alpha_i)\left(\frac{1}{\prod_{k=1}^{i-1}(2\cos\alpha_k)}\right)^2. \end{aligned}$$

We note that there are 2^{2i-1} strings of length s_i, 2^{2i-1} strings of length s_{vi}, and 2^{2n} bars of length ℓ_n. We summarize these mass results as follows.

Let a bar of length ℓ_0 and mass $m(\ell_0)$ buckle at the critical load $f(\ell_0)$. Replace this bar with a four-bar, four-tendon system called a *D-Bar* unit, as in Figure 3.1, designed to fail at the same external load $f(\ell_0)$. Replace each new bar with another *D-Bar* system. Repeat this self-similar process n times, where at each iteration there is freedom to choose angles α_i and the tension $t(s_i)$. However, we have chosen $t(s_i)$ to hold the same configuration of the structure in the absence of an external load. Then if each new structure is designed to fail (bars buckle and strings yield) at the same external force $f(\ell_0)$, then the mass of the new self-similar structure after the nth iteration is m_n, where $m_n = \mu_n m(\ell_0)$, and

$$\begin{aligned} \mu_n &= \left[2^{2n}m(\ell_n) + \sum_{i=1}^{n} 2^{2i-1}(m(s_i) + m(s_{vi}))\right]/m(\ell_0) \\ &= 2^{2n}\left(\frac{1}{\prod_{i=1}^{n}(2\cos\alpha_i)}\right)^{5/2} + \epsilon\sum_{i=1}^{n}\frac{2^{2i-2}(1+\tan^2\alpha_i)}{[\prod_{i=1}^{i-1}(2\cos\alpha_i)]^2}. \end{aligned} \tag{3.40}$$

Now consider the special case when $\alpha_i = \alpha$. In this case (3.40) yields

$$\begin{aligned} \mu_n &= 2^{2n}(2\cos\alpha)^{-5n/2} + \epsilon\sum_{i=1}^{n} 2^{2i-2}(1+\tan^2\alpha)(2\cos\alpha)^{-2(i-1)} \\ &= (2\cos^5\alpha)^{-n/2} + \epsilon\,(\cos^{-2n}\alpha - 1)/\sin^2\alpha, \end{aligned} \tag{3.41}$$

where we have used (3.18) for $\beta = \cos^{-2}\alpha$, $m = 1$.

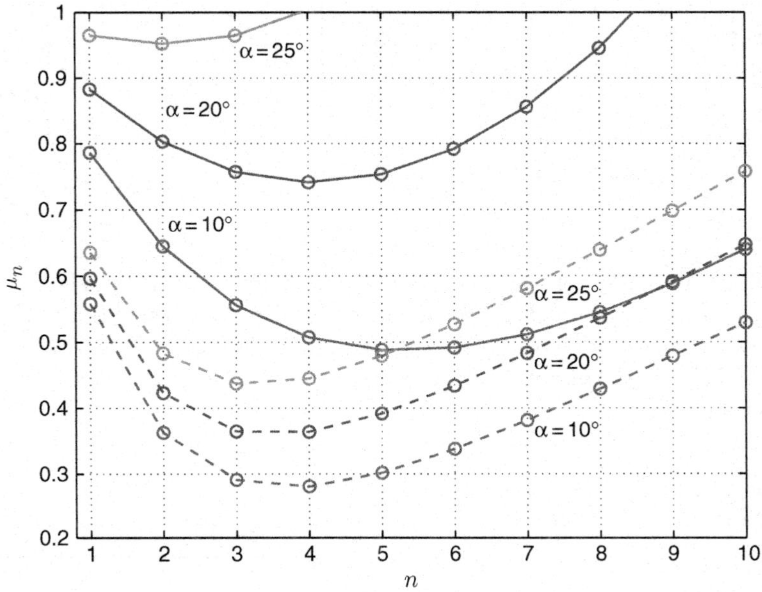

Figure 3.20: *D-Bar* unit: μ_n versus n, with $\epsilon = 0.05$ (*D-Bar, solid lines*), from (3.41). Also shown is μ_n versus n for the same ϵ (*T-Bar, dashed lines*), from (3.17)

A plot of $\mu_n(\alpha_i = \alpha)$ versus n appears in Figure 3.20. As n approaches infinity, tendon mass approaches infinity and the bar mass approaches zero. Clearly there exists a minimum mass at a finite number of iterations. It is straightforward, as in previous cases, to find the optimal number of iterations to get minimal mass. Hence the optimal complexity required for minimal mass is obtained by differentiating (3.41) with respect to n, set the expression to zero, and solve for n to get n^*, satisfying,

$$\left(\sqrt{\frac{1}{2\cos\alpha}}\right)^{n^*} = -\frac{4\epsilon\sin^2\alpha(\ln\cos\alpha)}{\ln 2 + 5(\ln\cos\alpha)}. \tag{3.42}$$

Or, explicitly,

$$n^* = \lfloor \frac{\ln(-4\epsilon\sin^2\alpha(\ln\cos\alpha)) - \ln(\ln(2\cos^5))}{\ln((2\cos\alpha)^{-1/2})} \rfloor. \tag{3.43}$$

We note from this equation that n^* approaches infinity as ϵ approaches zero, suggesting that with massless strings the mass of the set of bars approaches zero as the complexity n approaches infinity. This yields what we label as *D-Bar tensegrity fractals*, where in this case, the iterations $i = 1, 2, 3, \ldots$ in Figure 3.17 continue to infinity. The *D-Bar tensegrity fractals* fill the internal space of the structure as the configuration becomes arbitrarily

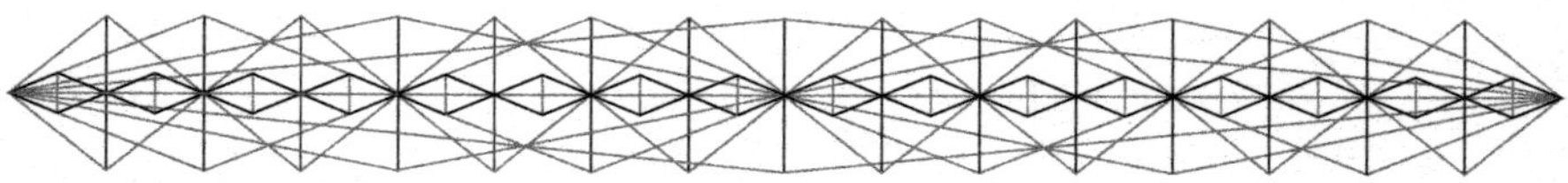

Figure 3.21: Optimal constant-width *T-Bar* column $\ell_0/w = 10$, $n^* = 5$, and $k = 4$, with last iteration using *D-Bar* units

dense. It is interesting to note that, in addition to filling space with a rule for self-similar iterations, a specific mechanical property (mass, strength, and stiffness) can be assigned to the structure at each iteration. Standard fractal theories from topology of course fill the space, but do not address system mechanical properties, given any material properties of the members of the topology.

Example 3.11

See Figure 3.17 for cases of $n = 3, 4, 5, 6$.

Example 3.12

From Figure 3.20 the optimal complexity is $n^* = 4$ for the system with $\epsilon = 0.05$ and $\alpha = 10°$ or $20°$ and $n^* = 3$ if $\alpha = 25°$.

Example 3.13

There are important reasons to combine *D-Bar* and *T-Bar* self-similar systems. It is easy to collapse a *D-Bar* unit by controlling the string s_{v1}, while the collapse of the *T-Bar* unit requires a folding procedure which seems more complex. Yet the *T-Bar* self-similar system can reduce more mass on each iteration than the *D-Bar* system. To combine the advantages of both, one can use the *T-Bar* self-similar iteration except on the last iteration. The last iteration will employ the *D-Bar* units. Figure 3.21 illustrates a deployable column design, where the column diameter is chosen to be constant.

3.3.3 Yielding in *D-Bar* Self-Similar Systems

As in the section on yielding in *T-Bar* self-similar systems, one can compute the number of iterations required for material yielding for *D-Bar* systems. We consider only the case with constant α. Equating the force required to yield and the force required to buckle yields the formula

$$(2\cos\alpha)^n = \frac{4\sigma_b^2}{\pi E}\frac{\ell_0^2}{f(\ell_0)}. \tag{3.44}$$

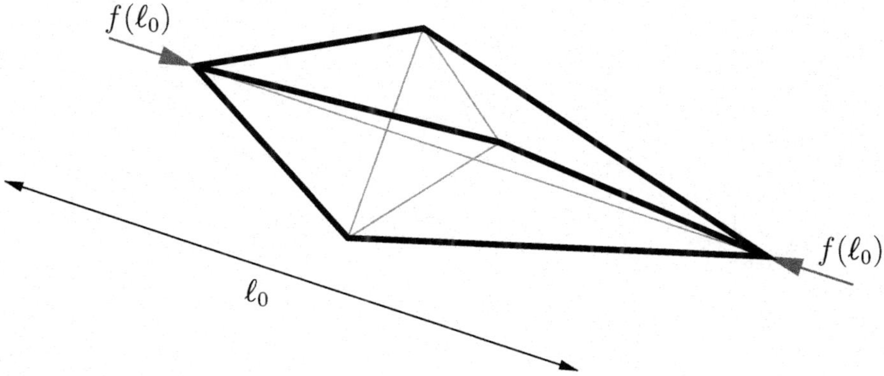

Figure 3.22: A three-dimensional *D-Bar* unit

Example 3.14

For a *D-Bar* column with constant $\alpha = \pi/18$, all steel members, with $\ell_0^2/f(\ell_0) = 1$, show that the number of iterations required to yield is given by

$$n = \frac{\ln(\ell_0^2/f(\ell_0)) + \ln(2\sigma)^2/\pi E}{\ln(2\cos\alpha)} = \frac{\ln(2\sigma)^2/\pi E}{\ln(2\cos\alpha)} = 22. \qquad (3.45)$$

The examples in the planar studies indicate that failure by buckling occurs with much smaller n, hence buckling is the expected mode of failure for *D-Bar* systems.

3.3.4 Three-Dimensional *D-Bar* System

See Figure 3.22 for the configuration of the three-dimensional *D-Bar* tensegrity unit. Since the mathematical details follow the same arguments as in the planar case, we rely on the previous sections to provide the step-by-step derivations. In this section we shall briefly present the results for this three-dimensional case.

Using the same arguments as in the planar case, we choose the prestress to yield the same geometry as the loaded structure. Assuming $\alpha_i = \alpha$ for all i, the relevant forces and lengths are given by

$$s_i = \ell_0/[2(2\cos\alpha)^{i-1}], \qquad s_{vi} = 2\,\ell_0\cos[\pi/(2N)]\sin\alpha/(2\cos\alpha)^i,$$
$$\bar{t}(s_i) = f(\ell_{i-1}), \qquad t(s_{vi}) = \bar{t}(s_{vi}) = f(\ell_i)\sin\alpha/\cos[\pi/(2N)],$$
$$f(\ell_i) = f(\ell_0)/(N\cos\alpha)^i, \qquad t(s_i) = 0,$$

where, as in Section 3.2.5, N refers to the number of bars at the node of application of the compressive force. In Figure 3.22, $N = 3$. After the usual

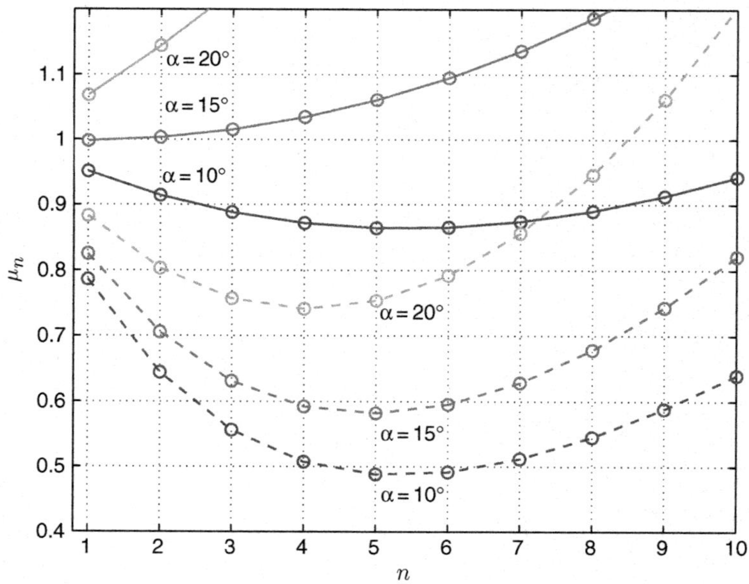

Figure 3.23: *D-Bar* unit: μ_n versus n, with $\epsilon = 0.05$ for planar ($N = 2$, *dashed lines*) and three-dimensional ($N = 3$, *solid lines*) columns, from (3.46) and (3.41)

algebraic manipulations we are led to the final result for the three-dimensional *D-Bar* system with constant α,

$$\mu_n^N = \left(\frac{N}{4\cos^5\alpha}\right)^{n/2} + \epsilon\,(\cos^{-2n}\alpha - 1)/\sin^2\alpha. \tag{3.46}$$

Note that the component associated with the string mass does not depend on N, and hence it is exactly the same as computed in the planar case in (3.41).

For example, substituting $n = 1$ and $N = 3$ yields for a single three-dimensional unit with six bars,

$$\mu_1^3 = \left(\frac{3}{4\cos^5\alpha}\right)^{1/2} + \epsilon\,(1 + \tan^2\alpha). \tag{3.47}$$

Hence mass reduction in the three-dimensional *D-Bar* unit requires $4\cos^5\alpha > 3$, or equivalently, $\alpha < 19.25°$. The planar case required only $\alpha < 29.48°$.

A comparison of (3.47) with (3.37) and a comparison between (3.46) and (3.41) contrasts the planar and three-dimensional *D-Bar* columns, with constant α. A plot of (3.46) with $N = 3$ and (3.41) (μ_n versus n) compares the two in Figure 3.23.

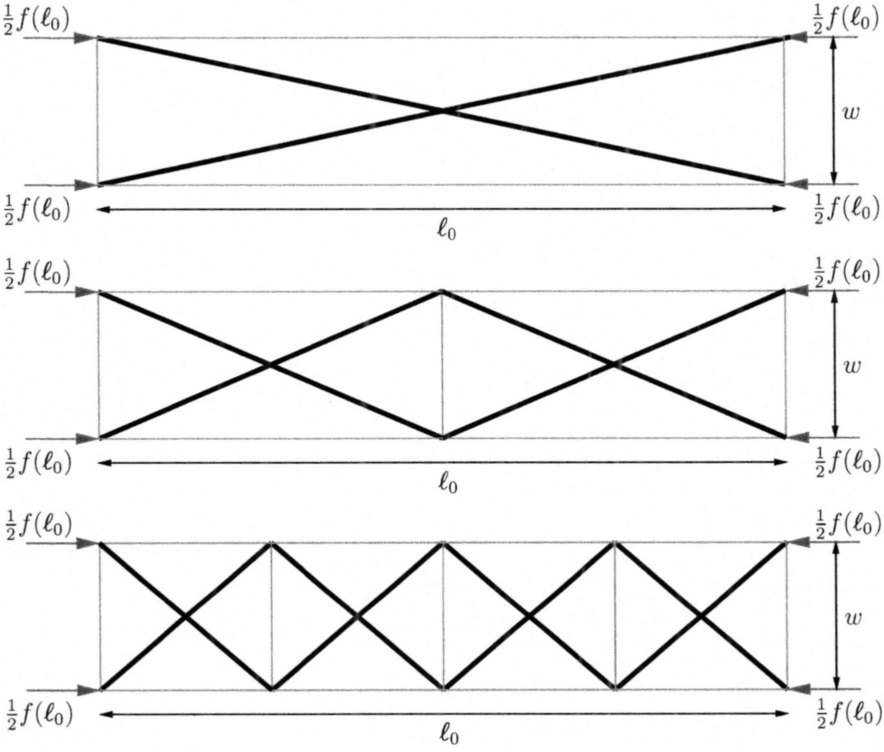

Figure 3.24: A USS *Box* structure under critical compressive load, $f(\ell_0)$

3.4 Unit-Self-Similar Designs

In the previous sections we replaced an *individual* compressive member by a more complex tensegrity structure. In this section we replace a specially chosen tensegrity unit (a subsystem, a collection of members) by a more complex set of units, while maintaining the outside dimensions of the structure. So if we were to label the iterative approach of the previous sections as a rule for *member-self-similar (MSS)* iterations, then we would label the rules in this section as *unit-self-similar (USS)* rules for iteration.

3.4.1 Using *Box* Units

We begin with a column example. As in the previous section for constant-width columns, consider a space of dimension $\ell_0 \times w$ that is available for constructing a compressive column. The first drawing in Figure 3.24 represents a single unit composed of two bars and four strings. This unit is a *Box* type of unit from Figure 3.2. The second drawing fills the same space with two *Box* units. Of course, when two strings are coincident, one can replace

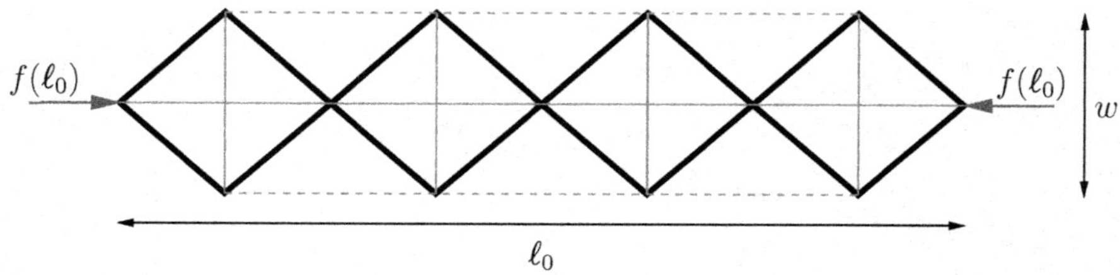

Figure 3.25: A USS *D-Bar* structure under critical compressive load, $f(\ell_0)$

them with a single string. The third drawing fills the same space with four *Box* units.

Consider a typical node, where the externally applied load is $f(\ell_0)$, the overall length and width of the beam is ℓ_0 and w, respectively, and if there are n units within the overall length ℓ_0, then the length of one unit is ℓ_0/n. Note also that $\tan\alpha_n = nw/\ell_0$. The sum of forces at the typical node yields,

$$f(\ell_n)\sin\alpha_n = t(s), \qquad f(\ell_0) = f(\ell_n)\cos\alpha_n,$$

and the mass is composed of $2n$ bars, leading to

$$\begin{aligned} m &= 2nc_b\ell_n^2\sqrt{f(\ell_n)} \\ &= 2nc_b[w^2 + (\ell_0/n)^2]\{f(\ell_0)^2[1 + (nw/\ell_0)^2]\}^{1/4}. \end{aligned}$$

Differentiating this mass expression with respect to n and setting this expression to zero yields, after a number of simplifications,

$$n = (\ell_0/w)\sqrt{2/3}. \tag{3.48}$$

Note that the optimal complexity of a column in compression, composed of USS iterations employing *Box* units, is a function only of the aspect ratio of the column, ℓ_0/w.

To summarize, we have shown that the assembly of n *Box* units as in Figure 3.24 to create a column of dimension $\ell_0 \times w$ has minimal mass if n is chosen to be $n = (\ell_0/w)\sqrt{2/3}$. Note that this is an example where the optimal complexity is finite.

3.4.2 Using *D-Bar* Units and *T-Bar* Units

Now consider the assembly of four *D-Bar* units, Figure 3.4, to form a column as in Figure 3.25. Since the units connect end to end at a point, there is no resistance to buckling of the column system, even though buckling of the individual units is avoided in the design. To prevent buckling of the column system we add strings on the column boundary, connecting the nodes on the width boundary with adjacent units, shown as dotted lines in Figure 3.25. That is, the ends of the strings s_{v1} of one unit are connected by a string to the end of the string s_{v1} of the adjacent unit. When this is done one can observe

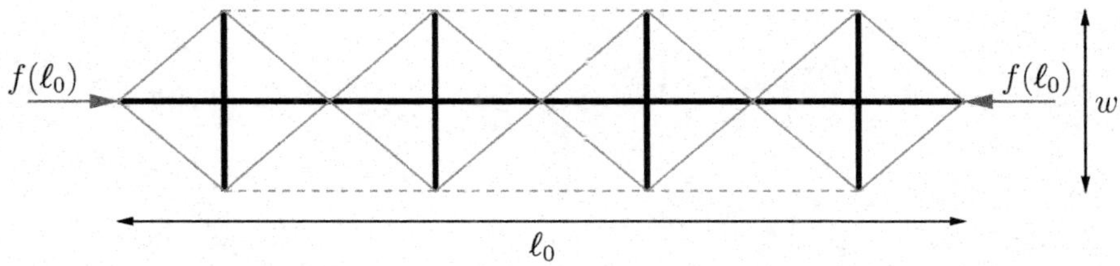

Figure 3.26: A USS *T-Bar* structure under critical compressive load, $f(\ell_0)$

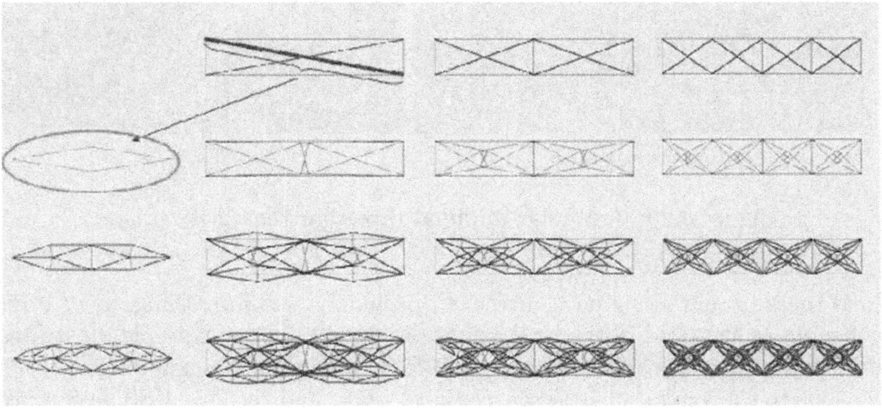

Figure 3.27: Member-self-similar (MSS) concepts can be applied on each compressive member (down the *columns* of the figure). Unit-self-similar (USS) concepts can be applied on each unit (along the *rows* of the figure). A minimal mass occurs along each direction, providing a minimal mass over both concepts as in the *lower right* design

that the topology of the system is very similar to the topology obtained by the interconnection of *Box* units, described above, with the exception that the ends of the column form a point (from the left and right ends of a *D-Bar* unit).

Finally, consider the assembly of four *T-Bar* units, Figure 3.4, to form a column as in Figure 3.26. One can see that without the stabilizing strings, the USS *D-Bar* and *T-Bar* are dual structures. It is interesting to note that the same set of stabilizing strings can be added to both structures even though they are dual.

We have shown how to minimize mass using self-similar iterations of the members (MSS), and we have shown how to minimize mass using the units (USS). Figure 3.27 illustrates both concepts in the planar case of a column. Member-self-similar (MSS) concepts can be applied on each compressive member (down the columns of the figure). Unit-self-similar (USS) concepts can be applied on each unit (along the rows of the figure). A min-

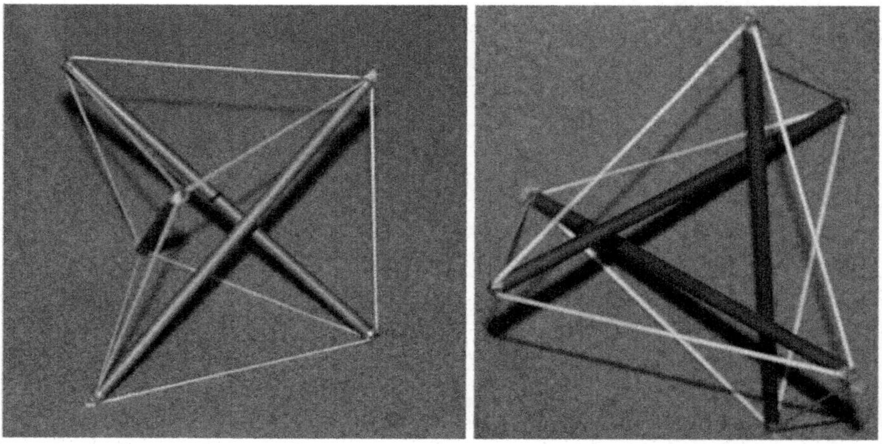

(a) Perspective (b) Top View

Figure 3.28: A regular minimal three-bar tensegrity prism

imal mass occurs along each direction, providing a minimal mass over both concepts as sketched in the lower right design, where the example illustrates only these concepts and does not provide numbers for the example.

Due to the similarity between *D-Bar*, *T-Bar*, and the *Box* USS design, we will not examine the *D-Bar* design further. Instead, we move on to a more interesting class of self-similar structures that are natural extensions of the *Box* design in three dimensions. These structures will be based on *tensegrity prisms*, which we study in detail in the next section.

3.5 Tensegrity Prisms

Tensegrity prisms have been introduced in Section 1.4.1 along with some nomenclature regarding regularity (whether the top and bottom polygons are regular and lie in parallel planes, potentially with different radius) and minimality (whether the prism has the minimum possible number of strings for stable equilibrium). The prism in Figure 3.28 is both regular and minimal. This is the first class of prisms we will study.

3.6 Minimal Regular Prisms

Minimal regular tensegrity prisms can have any number of bars $p \geq 2$. *Regular* p-bar tensegrity prisms have top and bottom regular p-polygons lying in parallel planes. The radius of the top and bottom polygons can be different. In a regular p-bar prism, the minimum number of strings necessary for stable equilibrium is $3p$, namely p *top strings*, p *bottom strings*, and p

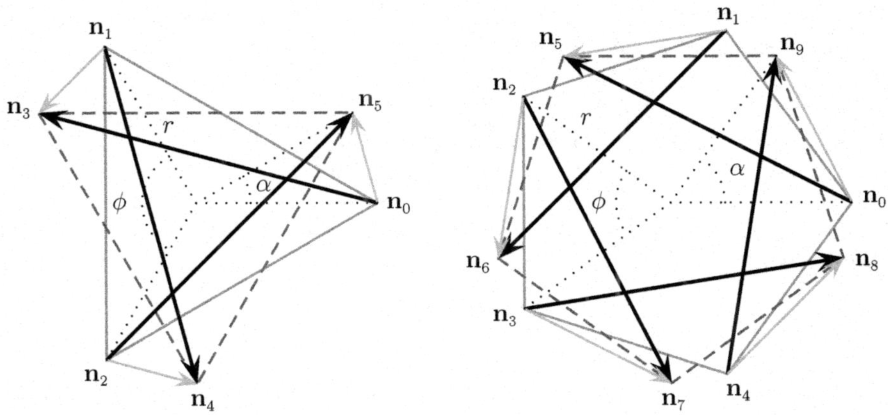

Figure 3.29: Top view of minimal regular tensegrity prisms (with $r_b = r_t = r$)

vertical strings. Such strings can be visualized in Figure 3.28 for the case of a three-bar minimal regular prism. A p-bar prism with $3p$ strings is a minimal prism even if it is not a regular prism. In the following section we discuss properties of minimal regular tensegrity prisms.

3.6.1 Equilibrium

Let r_t and r_b denote the radius of the top and bottom regular polygons of a p-bar regular tensegrity prism. The distance separating these two parallel polygons, sometimes called the *height* of the prism, is denoted by h. The *twist angle*, that is, the angle formed by the projection of the top and bottom polygons, is denoted by α. The angle ϕ is the characteristic angle of the regular polygons, that is, $\phi = 2\pi/p$. Most of these quantities are depicted in Figure 3.29, where $r_t = r_b = r$ for $p = 3$ and $p = 5$. As mentioned before, minimal tensegrity prisms have p bars and $3p$ strings. In Figure 3.29 the p bars are shown in black, e.g., $\mathbf{n}_4 - \mathbf{n}_1$ in Figure 3.29(a), p top strings are shown in *dotted blue*, e.g., $\mathbf{n}_4 - \mathbf{n}_3$ in Figure 3.29(a), and p bottom strings in *solid red*, e.g., $\mathbf{n}_2 - \mathbf{n}_1$ in Figure 3.29(a). Arrows indicate elements that connect to the bottom and top planes (the head of the arrow is at the top plane). The green members are p vertical strings, e.g., $\mathbf{n}_3 - \mathbf{n}_1$ in Figure 3.29(a).

We now present the fundamental relationship for equilibrium of self-stressed tensegrity prisms without any external loads. Detailed derivations are given at the end of the chapter.

Let γ_b, γ_t, and γ_v be the force densities, i.e., the ratio between the member force and the member length (see Section 2.2 for details) on all bottom, all top, and all vertical strings, respectively, and λ_b be the force density on all bars. The fact that all groups of strings and all bars share the same force coefficients follow from symmetry implied by the regularity of the prism. Without external forces, for equilibrium, we should have

$$\begin{pmatrix} \gamma_t \\ \gamma_b \\ \gamma_v \end{pmatrix} = \lambda_b \begin{pmatrix} \rho^{-1}(2\sin(\pi/p))^{-1} \\ \rho\,(2\sin(\pi/p))^{-1} \\ 1 \end{pmatrix}, \tag{3.49}$$

where $\rho := r_t/r_b$ is the ratio between the top and bottom radii. Note that the twist angle

$$\alpha = \frac{\pi}{2} - \frac{\pi}{p} \tag{3.50}$$

is uniquely defined for any p. For instance, for $p = 3$ then $\alpha = 30°$, $p = 4$ then $\alpha = 45°$, $p = 6$ then $\alpha = 60°$, which are well-known relationships [CB98].

The above formulas are scalable in the sense that λ_b and the unit geometry (h, r_b) or (h, r_t) can be chosen arbitrarily without affecting equilibrium. Increasing $\lambda_b > 0$ increases the overall level of prestress in the unit. Note that all γ's are positive if $\lambda_b > 0$. As for the geometry, changing h and say r_b will simply scale the force of the members to match the unit geometry. Also note that

$$\gamma_b = \rho^2 \gamma_t,$$

so that, as expected, the top and bottom strings will have the exact same forces when the top and bottom polygons have the exact same radii.

3.6.2 Design Under Compressive Load

Consider a regular minimal tensegrity prism subject to a total compressive load $f(\ell_0)$, as shown in Figure 3.30 for the case $p = 3$. We assume now that $r_t = r_b = r$, that is, that the top and bottom polygons are congruent. Note that in order to make the notation consistent with other sections in this chapter, the height h is now seen as the prism *length* and is therefore labeled ℓ_0. Likewise, the prism *width* is $w = 2r$. This will allow us to compare the results of this section with the other topologies we have studied.

Even though one can compute the exact forces on all members of a tensegrity prism in equilibrium under compressive load (see, for instance [Ske05]), here we will adopt the same "small displacements" paradigm used in the previous sections in order to design a tensegrity prism. As before, we assume that a stable regular minimal prism is to be designed such that when subject to a compressive load $f(\ell_0)$ equally distributed on the p top and p bottom nodes, the vertical component of the vertical string is zero. Under the small displacements assumption we have that

$$\mathbf{f}_v^z = \ell_0 \gamma_v = f(\ell_0)/p \qquad \Longrightarrow \qquad \gamma_v = f(\ell_0)/(\ell_0\, p).$$

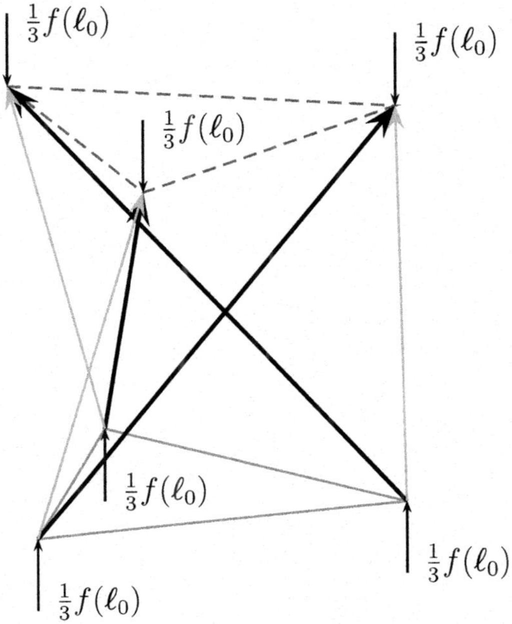

Figure 3.30: Minimal regular tensegrity prism under compressive load

This allows one to calculate all force densities in all members of the structure, namely

$$\lambda_b = \gamma_v = \frac{f(\ell_0)}{\ell_0\, p}, \qquad \gamma_t = \gamma_b = \frac{\lambda_b}{2\sin(\pi/p)} = \frac{f(\ell_0)}{2\;\ell_0\, p \sin(\pi/p)}.$$

After computing the lengths of the bars and the strings

$$\|\mathbf{b}\| = \sqrt{\ell_0^2 + 2r^2[1+\sin(\pi/p)]},$$

$$\|\mathbf{s}_v\| = \sqrt{\ell_0^2 + 2r^2[1-\sin(\pi/p)]}, \qquad \|\mathbf{s}_t\| = \|\mathbf{s}_b\| = 2r\sin(\pi/p),$$

we can compute the total mass needed for the bar to buckle at the given load $f(\ell_0)$ as

$$m_{b1} = p\, c_b \|\mathbf{b}\|^2 \sqrt{\lambda_b \|\mathbf{b}\|} = c_b \sqrt{\frac{p f(\ell_0)}{\ell_0}} \left(\ell_0^2 + 2r^2[1+\sin(\pi/p)]\right)^{5/4}.$$

As before, we are interested in mass savings as compared to a single bar under compression $m(\ell_0) = c_b \ell_0^2 \sqrt{f(\ell_0)}$. This is why we compute the mass savings ratio

$$\mu_{b1} = \frac{m_{b1}}{m(\ell_0)} = \frac{\sqrt{p}}{z^5} \left(\frac{1 + 2z^4 + \sin(\pi/p)}{2}\right)^{5/4},$$

where, as before, $z = \sqrt{\ell_0/w} = \sqrt{\ell_0/(2r)}$. As for the mass of the strings

$$m_{s1} = p\,c_s \left(\gamma_v\|\mathbf{s}_v\|^2 + \gamma_b\|\mathbf{s}_b\|^2 + \gamma_t\|\mathbf{s}_t\|^2\right) = \frac{c_s f(\ell_0)}{\ell_0}\left(\ell_0^2 + 2r^2(1+\sin(\pi/p))\right),$$

and

$$\mu_{s1} = \frac{m_{s1}}{m(\ell_0)} = \epsilon\left(1 + \frac{1+\sin(\pi/p)}{2z^4}\right),$$

where ϵ is as defined in (3.11). The total mass savings for a single minimal regular tensegrity p-bar prism are then

$$\begin{aligned} \mu_1 &= \mu_{b1} + \mu_{s1} \\ &= \sqrt{p}\left(\frac{1+2z^4+\sin(\pi/p)}{2}\right)^{5/4} + \epsilon\left(1 + \frac{1+\sin(\pi/p)}{2z^4}\right). \end{aligned} \tag{3.51}$$

Note that $\mu \geq \sqrt{p}\,(1+\sin(\pi/p))^{5/4} \geq \sqrt{2} > 1$, so that no mass savings are possible for a single unit. This will be overcome in the next section where a unit-self-similar design will be developed based on minimal regular tensegrity p-bar prisms.

3.7 Tensegrity Columns

Consider now the unit-self-similar structure in Figure 3.31 subject to a total compressive load $f(\ell_0)$. This is a class 2 tensegrity structure based on a regular minimal tensegrity prism. Because $r_t = r_b = r$, stable minimal regular p-bar prisms can be stacked as shown, ensuring stability of the overall structure. In this design, the length of each unit is $\ell_n = \ell_0/n$, where n is the total number of units.

3.7.1 Unit-Self-Similar Design

Defining $z = \sqrt{\ell_0/(2r)}$ and repeating the same operations as in the previous section, one obtains

$$\mu_n = \mu_{bn} + \mu_{sn}, \qquad \mu_{bn} = \frac{\sqrt{p}}{nz^5}\left(\frac{n^2+2z^4+n^2\sin(\pi/p)}{2}\right)^{5/4}, \tag{3.52}$$

$$\mu_{sn} = \epsilon\left(1 + \frac{n^2[1+\sin(\pi/p)]}{2z^4}\right). \tag{3.53}$$

Note that in this design the mass of the strings *decreases* monotonically with n but is bounded from below by ϵ. On the other hand, the mass of the bars has a distinct minimum. After differentiating μ_{bn} with respect to n, one finds

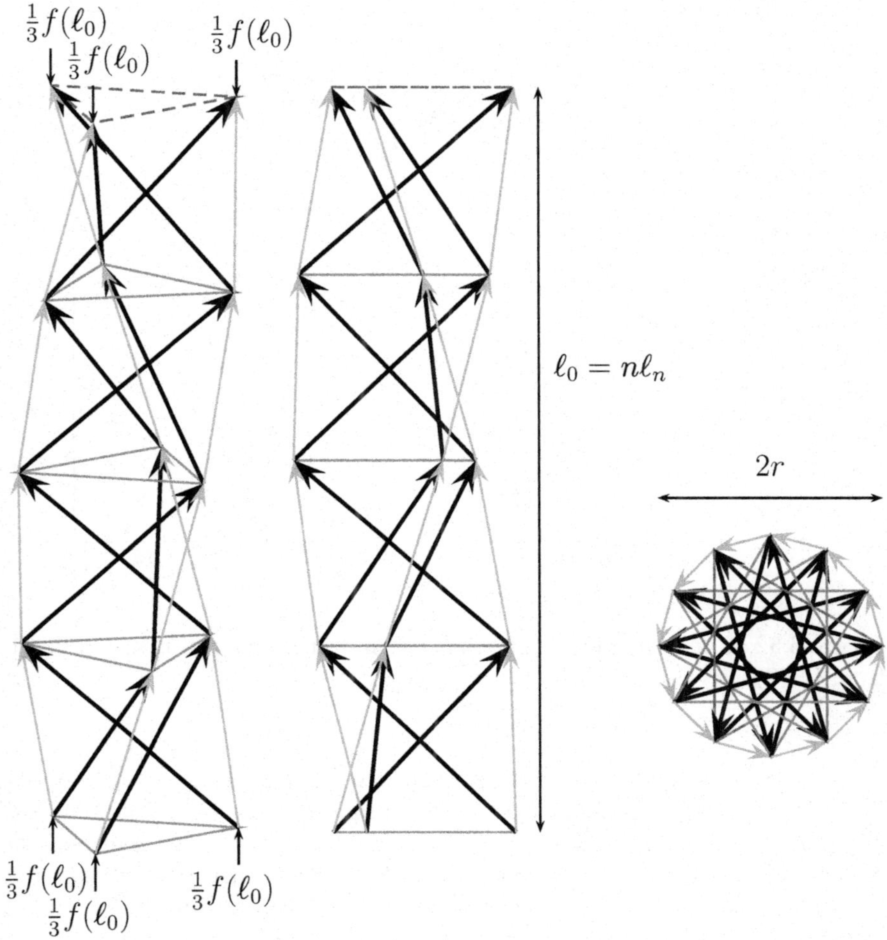

(a) Perspective view (b) Side view (c) Top view

Figure 3.31: Unit-self-similar tensegrity column under compressive load

that the number of units that minimizes μ_{bn} is

$$n^* = \left\lfloor \frac{2z^2}{\sqrt{3[1+\sin(\pi/p)]}} \right\rfloor = \left\lfloor \frac{2}{\sqrt{3(1+\sin(\pi/p))}} \frac{\ell_0}{w} \right\rfloor . \tag{3.54}$$

These formulas are illustrated by the following examples.

Example 3.15

The plot in Figure 3.32 shows μ_n as a function of n for a minimal regular three-bar prism, plotted for various values of ℓ_0/w and $\epsilon = 0.001$. Note that, as

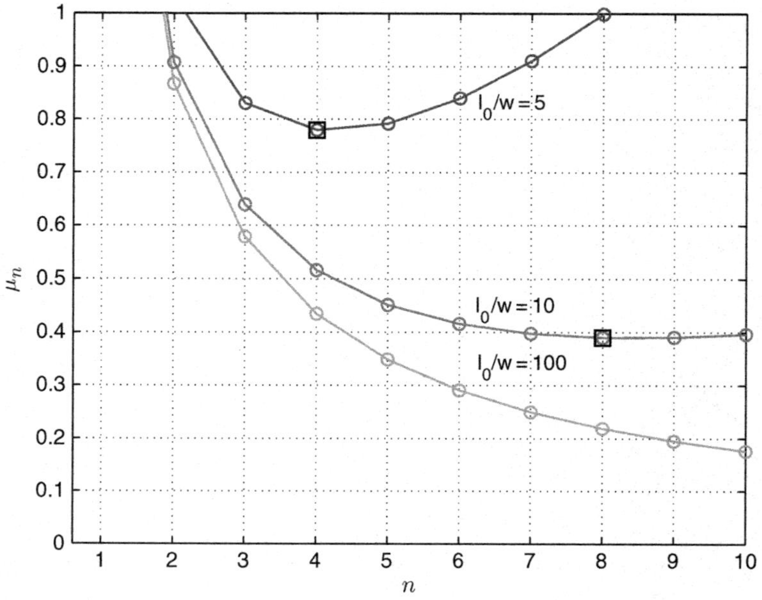

Figure 3.32: Unit-self-similar three-bar prism design: μ_n versus n, $\epsilon = 0.001$, from (3.52)

expected from the discussion in the previous section, one needs $n \geq 2$ to obtain some mass savings. The points marked with a square indicate the global minima. The global minimum for $\ell_0/w = 100$ is $\mu_n = 0.0405$ at $n = 82$. Whereas the savings are comparable to those obtained with say, the planar *T-Bar* design, this is achieved at a much higher number of units. The mass savings for smaller values of ℓ_0/w occur for smaller n but are not as impressive. Note that

$$n^*(\ell_0/w) = \{4, 8, 84\}, \qquad \ell_0/w = \{5, 10, 100\},$$

which indicates that n^* seems to be a reasonably accurate estimate of the global optimum.

Example 3.16

The plot in Figure 3.33 shows μ_n as a function of n for a minimal regular three-bar prism, plotted for various values of ϵ. In the case $\epsilon = 0.05$, n^* is 8 which again turned out to be an accurate estimate of the global minima.

From the previous example, n^* seems to be an accurate estimate of the global minima. It is also easily related to ℓ_0/w. Motivated by this we compute an upper bound to the optimal mass gain by substituting n^* in (3.52) in order

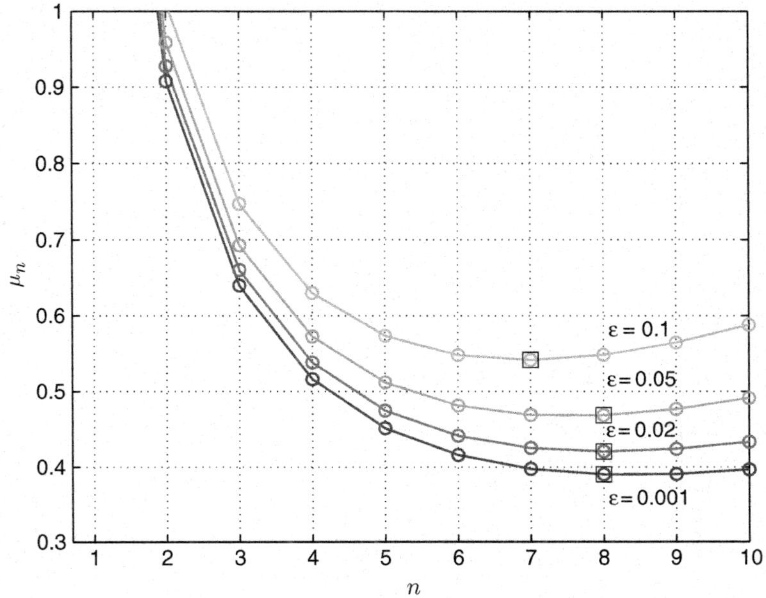

Figure 3.33: Unit-self-similar three-bar prism design: μ_n versus n, $\ell_0/w = 10$, from (3.52)

to obtain

$$\mu_{n^*} = \frac{1}{\ell_0/w} \frac{5 \times 15^{1/4}}{6} \sqrt{p[1 + \sin(\pi/p)]} + \epsilon \frac{10}{6}. \quad (3.55)$$

The above formula is remarkable and shows that the optimal mass gain is linear in ϵ and that the component of the mass due to the strings (the term multiplied by ϵ) is independent of the number of bars per prism p. Also, for a given p, the optimal mass gain is inversely proportional to the aspect ratio ℓ_0/w.

Example 3.17

The plot in Figure 3.34 shows μ_{n^*} as a function of ℓ_0/w for a minimal regular p-bar prism, plotted for various values of p. Note by the boxed points in the figure that for a given p a certain minimum aspect ratio is needed for mass savings.

3.8 Tensegrity Plates

Some regular p-bar tensegrity prisms can be connected in such a way as to generate plates. However, only a few choices of p, namely $p = \{3, 4, 6\}$,

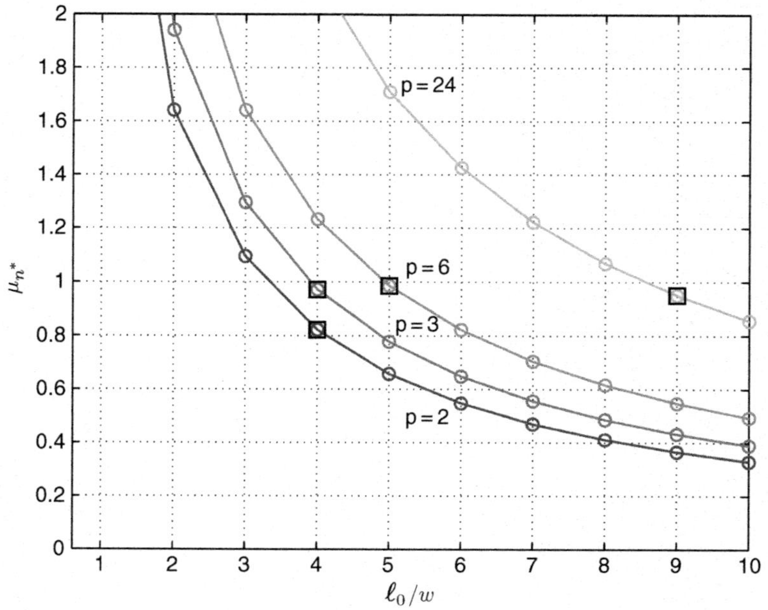

Figure 3.34: Unit-self-similar p-bar prism design: μ_{n^*} versus ℓ_0/w, from (3.55)

generate plates in which all top (and bottom) nodes lie on the same plane. In this chapter we shall limit ourselves to discuss flat plates assembled using three-bar minimal tensegrity prisms. This can be done in two different ways, which we call topology A and topology B.

3.8.1 Topology A

In this plate topology two three-bar minimal regular tensegrity prisms of the same height h and the same top and bottom radius $r_t = r_b = r$ are placed such that two of the nodes of the second prism, one at the top and one at the bottom, lie exactly on one top and one bottom string of the first prism. The idea is illustrated in the top view shown in Figure 3.35, where the nodes of the second prism are labeled $\mathbf{n}'_i$, $i = 0, \ldots, 5$. The figure also suggests a solution to the problem of characterizing the geometry of such plates. Indeed, this geometry is entirely determined by looking at the top (or bottom) projection of the prism's two parallel planes. For this reason the next derivations are done with two-dimensional vectors, with no z-component.

We look for a two-dimensional translation vector $\mathbf{t}_{3A}$ such that the bottom node $\mathbf{n}'_0$ of the second prism lies on the line formed by the nodes $\mathbf{n}_1$ and $\mathbf{n}_2$ of the first prism. At the same time the top node $\mathbf{n}'_5$ lies on the line formed by the top nodes $\mathbf{n}_3$ and $\mathbf{n}_4$. That is, we look for a vector $\mathbf{t}_{3A}$ and scalars

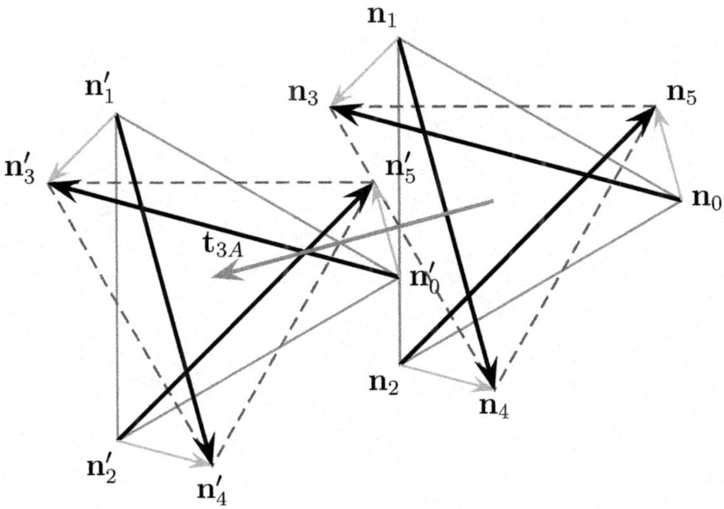

Figure 3.35: Three-bar plate connection – topology A – top view

$0 \leq \beta_b \leq 1$ and $0 \leq \beta_t \leq 1$ such that

$$\mathbf{n}'_0 = \mathbf{t}_{3A} + \mathbf{n}_0 = \beta_b\, \mathbf{n}_2 + (1-\beta_b)\mathbf{n}_1, \tag{3.56}$$

$$\mathbf{n}'_5 = \mathbf{t}_{3A} + \mathbf{n}_5 = \beta_t\, \mathbf{n}_3 + (1-\beta_t)\mathbf{n}_4. \tag{3.57}$$

The answer to the above linear algebra problem, shown at the end of the chapter, is of the form

$$\begin{bmatrix} \beta_b \\ \beta_t \end{bmatrix} = (\sqrt{3}-1)\begin{bmatrix} 1 \\ 1 \end{bmatrix}, \qquad \mathbf{t}_{3A} = \begin{bmatrix} -1 \\ \sqrt{3}-2 \end{bmatrix}\frac{3r}{2}. \tag{3.58}$$

Due to the symmetry of the problem, a prism can be surrounded by six other prisms which are translated by

$$\mathbf{t}_{3Aj} = R^j_{\pi/3}\mathbf{t}_{3A}, \qquad j = 0, \ldots, 5,$$

generating the pattern shown in Figure 3.36. Note that a minimum of three prisms, e.g., the central unit and $j = k, \ldots, k+1$ for any k, are required to connect in order to obtain a stable plate. This is due to the fact that only two nodes are connected between two prisms, which in three dimensions leaves 1 out of 3 degrees of freedom unstabilized. Note that with three prisms we have six linearly independent node connections.

3.8.2 Topology B

In this topology two three-bar minimal regular tensegrity prisms of the same height h and the same top and bottom radius $r_t = r_b = r$ are placed such that one node of the first prism lies exactly on a string of the second prism and one

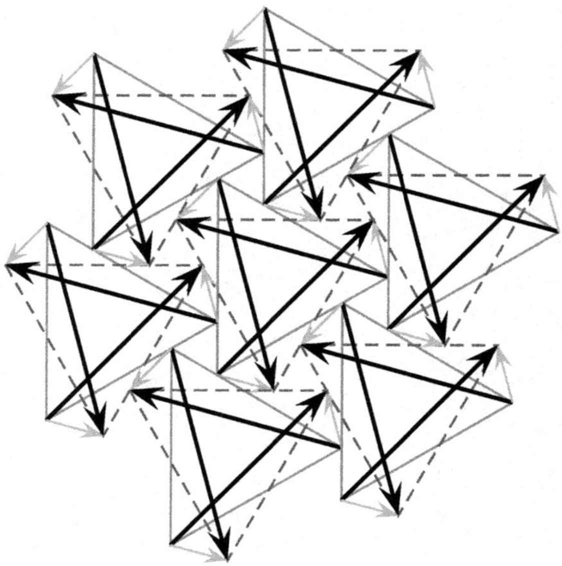

Figure 3.36: Three-bar plate – topology A – top view

node of the second prism lies exactly on a string of the first prism. The idea is illustrated in Figure 3.37. As for topology A, we look for a two-dimensional translation vector $\mathbf{t}_{3B}$ such that the bottom node $\mathbf{n}_0'$ of the second prism lies on the line formed by the nodes $\mathbf{n}_1$ and $\mathbf{n}_2$ of the first prism. In addition, the top node $\mathbf{n}_3$ of the first prism should lie on the line formed by the top nodes $\mathbf{n}_4'$ and $\mathbf{n}_5'$. That is, we look for a vector $\mathbf{t}_{3B}$ and scalars $0 \le \beta_b \le 1$ and $0 \le \beta_t \le 1$ such that

$$\mathbf{n}_0' = \mathbf{t}_{3B} + \mathbf{n}_0 = \beta_b\, \mathbf{n}_2 + (1 - \beta_b)\mathbf{n}_1,$$

which is the same equation as in (3.56), and

$$\begin{aligned} \mathbf{n}_3 &= \beta_t\, \mathbf{n}_4' + (1 - \beta_t)\mathbf{n}_5', && (3.59)\\ &= \beta_t\, (\mathbf{n}_4 + \mathbf{t}_{3B}) + (1 - \beta_t)(\mathbf{n}_5 + \mathbf{t}_{3B}), \\ &= \mathbf{t}_{3B} + \beta_t \mathbf{n}_4 + (1 - \beta_t)\mathbf{n}_5. && (3.60)\end{aligned}$$

Proceeding as in the case of topology A, the solution

$$\begin{bmatrix}\beta_b\\ \beta_t\end{bmatrix} = (2-\sqrt{3})\begin{bmatrix}1\\1\end{bmatrix}, \qquad \mathbf{t}_{3B} = \begin{bmatrix}-1\\ 2-\sqrt{3}\end{bmatrix}\frac{3r}{2} \qquad (3.61)$$

is derived at the end of the chapter.

As in the previous section, a unit can be surrounded by six other units which are translated by

$$\mathbf{t}_{3Bj} = R_{\pi/3}^{j}\mathbf{t}_{3B}, \qquad j = 0, \ldots, 5,$$

generating the pattern shown in Figure 3.38.

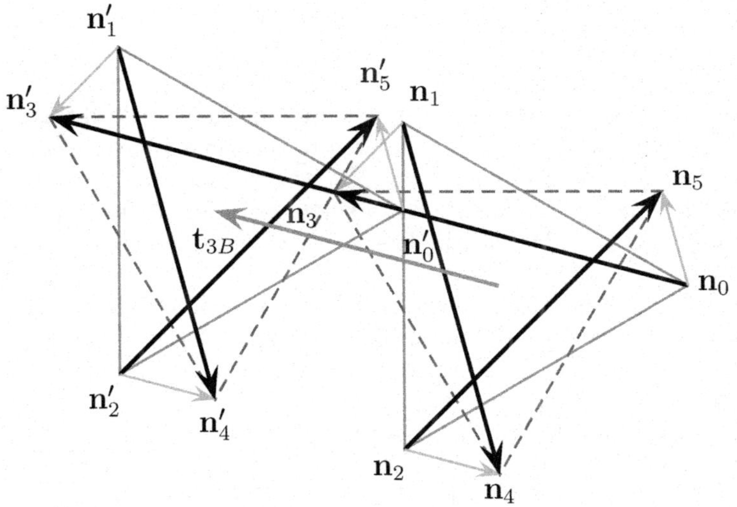

Figure 3.37: Three-bar plate connection – topology B – top view

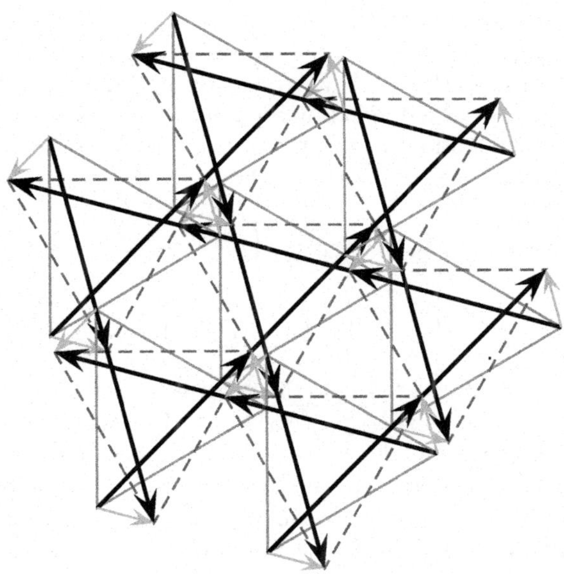

Figure 3.38: Three-bar plate – topology B – top view

3.8.3 Design Under Compressive Load

Because $r_t = r_b = r$, we shall be able to borrow much of the derivations in Section 3.6.2. This is indeed the case after noticing that for both topologies A and B all prisms are connected by simply "sliding" one prism close to another and that the static equilibrium of the individual prisms remains unaffected by such connections.

Indeed, for an individual three-bar prism in a plate subject to a compressive load $f(\ell_0)$ equally distributed among its three top and three bottom nodes, the total mass saving as compared to the same force applied to a single bar is given by (3.51), where $z = \sqrt{\ell_0/w}$ is the square root of the aspect ratio and ϵ is defined by (3.11). For a single prism or radius r, the width is taken simply as $w = 2r$, and the length is equal to the height of the prism, i.e., $\ell_0 = h$.

For unit-self-similar columns as in Section 3.7, the width $w = 2r$ and the total length ℓ_0 remain constant while the height of the individual units is reduced to $h_n = \ell_n = \ell_0/n$. Also total length ℓ_0 is larger than width w, so that $z = \sqrt{\ell_0/w} > 1$.

Let us now consider that a tensegrity plate with width ℓ_0 is to be built using topologies A and B to cover an area A. For that sake we use n units of height $h = \ell_0$ connected by topologies A or B. Then the area of each unit is πr^2 and the total area

$$A = \xi n \pi r^2,$$

where $\xi \in [0, 1]$ is a parameter that accounts for the overlap between the areas of the units in either topologies (more on the determination of ξ later). From this definition we have that $r = \sqrt{A/\xi n \pi}$. By noting that the force in each individual unit is now also divided by the number of units one can repeat the steps in Sections 3.6.2 and 3.7 with $\mathbf{f}_v^z = f(\ell_0)/(3n)$ and the definition of the aspect ratio parameter $z^2 = \sqrt{A}/\ell_0$ to obtain

$$\mu_n = \mu_{bn} + \mu_{sn}, \qquad \mu_{bn} = \frac{\sqrt{3}}{n^{3/4}\pi^{5/4}} \left(n\pi + \xi^{-1}(2+\sqrt{3})z^4\right)^{5/4}, \tag{3.62}$$

$$\mu_{sn} = \epsilon \left(1 + \frac{2+\sqrt{3}}{n\pi\xi} z^4\right). \tag{3.63}$$

Several remarks are in order. First, the quantity μ_n still expresses the total mass normalized by $m(\ell_0) = c_b \ell_0^2 \sqrt{f(\ell_0)}$ as in (3.7), where $m(\ell_0)$ is the minimal mass of a single bar under a lumped compressive load $f(\ell_0)$. However, this quantity is not meaningful in the case studied here, in which the force is distributed over the area A. The reason for using the same normalization is to be able to work with the non-dimensional parameters z and ϵ, whereas z has been properly redefined to be consistent with the previously used definitions which are related to the square root of the "aspect ratio". The consequence is that when looking for the plots of μ_n one should not draw any comparison with unity.

As in the previous section, the mass of the strings decreases monotonically with n and is bounded from below by ϵ whereas the mass of the bars has a distinct minimum. As in the previous section, one may try to compute this minimum with respect to n by ignoring the string mass. The task is, however,

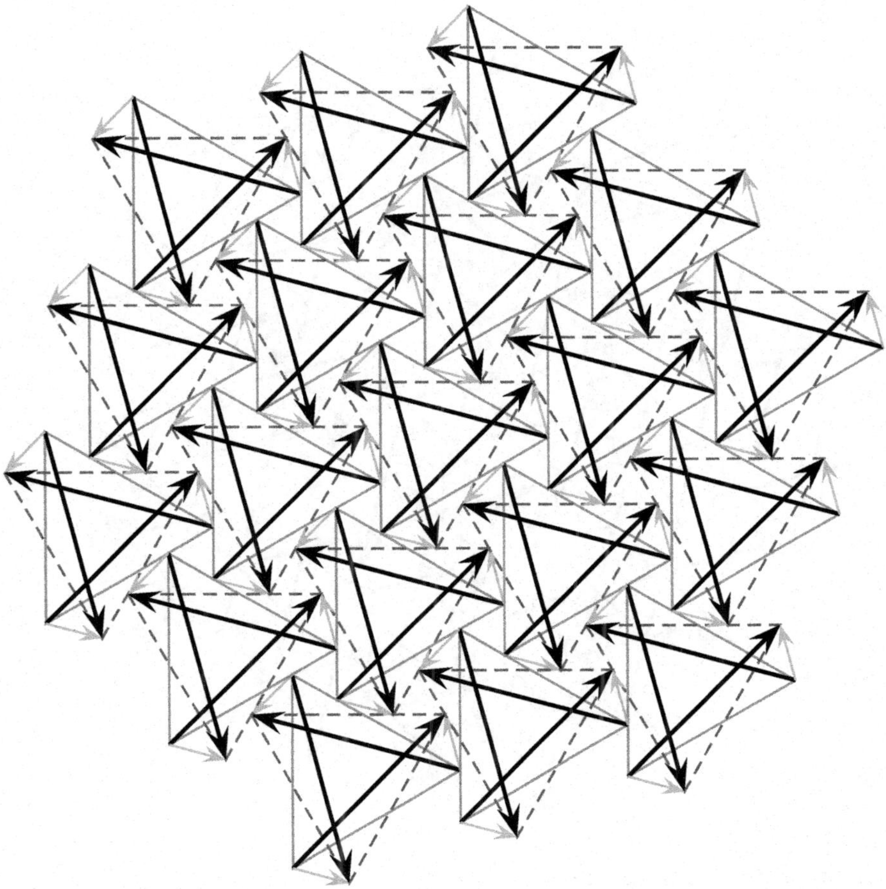

Figure 3.39: Three-bar hexagonal plate with three rings – topology A – top view

more involved since ξ is often a function of n as well. The next section will provide a concrete example.

3.8.4 Hexagonal Three-Bar Flat Plates

Consider hexagonal flat tensegrity plates, such as the ones in Figure 3.36 built using topology A. Such plates can be extended indefinitely by attaching more "rings" to the sides of an existing plate. For example, Figure 3.36 has two rings, whereas Figures 3.39 and 3.40 show plates built with topology A with three and four rings, respectively. Similar plates can be constructed using topology B, as in Figures 3.38 and 3.41 with two and three rings, respectively.

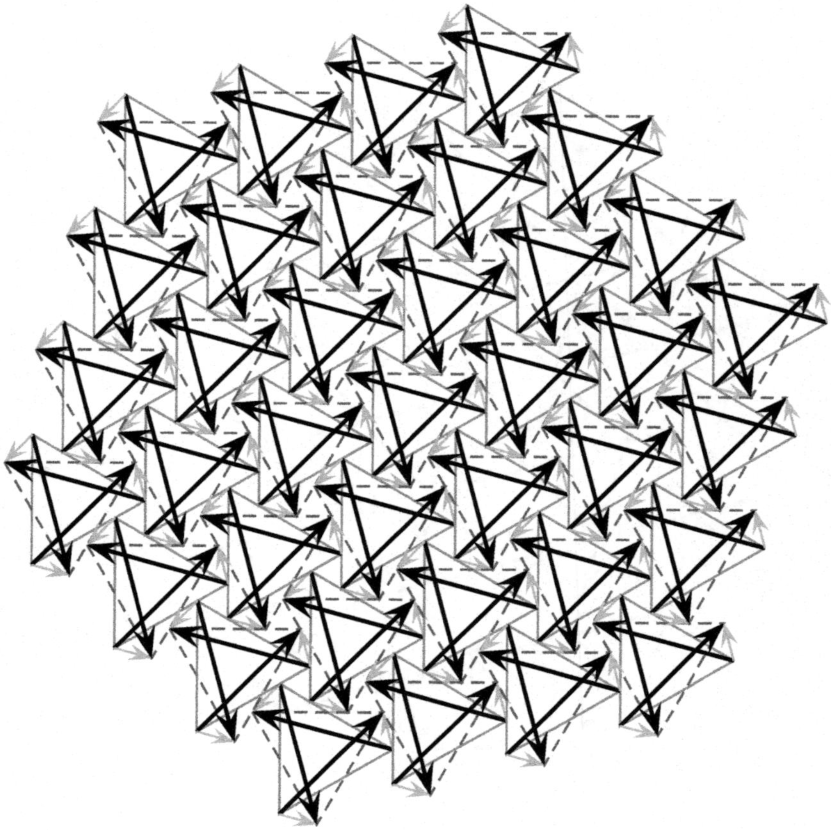

Figure 3.40: Three-bar hexagonal plate with four rings – topology A – top view

For such plates we can compute the value of ξ exactly. Indeed, if the individual units have radius r, hence area $a = 3r^2\cos(\pi/6)$, then the total area of these plates is $A = 3R^2\cos(\pi/6)$. The radius R is computed as

$$R = \left(\frac{1}{2} + \|t_3\|(n_r - 1)\right) r, \qquad \|t_3\| := 3\sqrt{2-\sqrt{3}},$$

where $n_r > 1$ is the number of rings and $\|t_3\| = \|t_{3A}\| = \|t_{3B}\|$ is the center-to-center distance computed previously. Note that this is the same value for both topologies A and B even though $t_{3A} \neq t_{3B}$! With that and the fact that the number of prisms in a hexagonal plate is

$$n = 1 + 3n_r(n_r - 1), \tag{3.64}$$

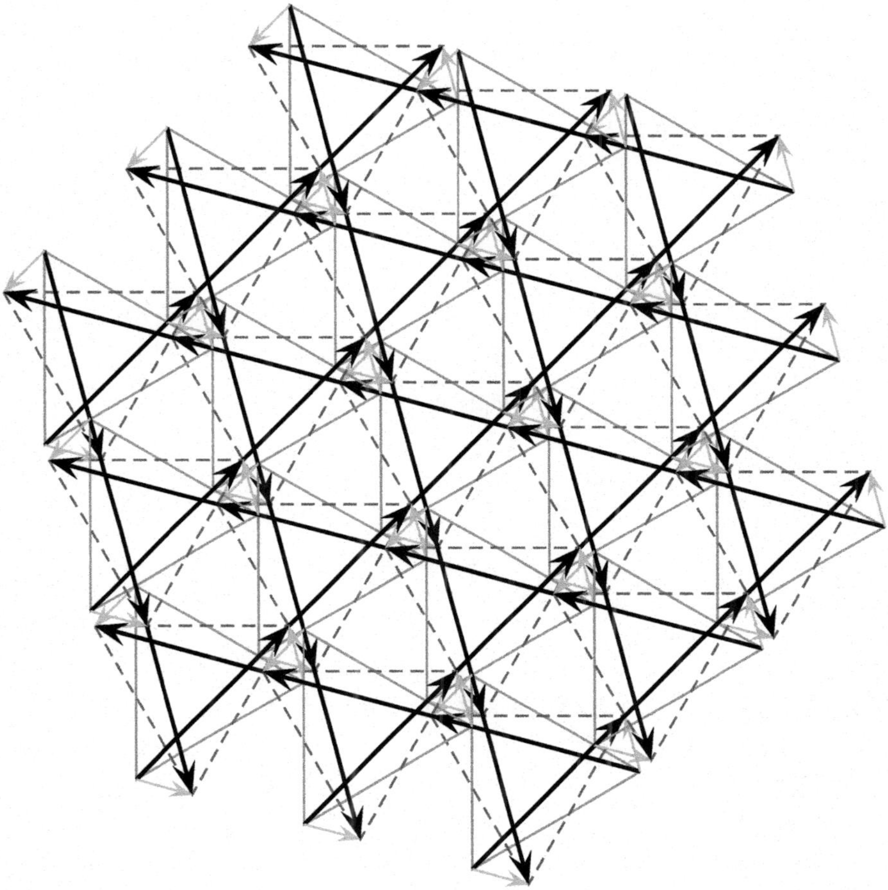

Figure 3.41: Three-bar hexagonal plate with three rings – topology B – top view

one can compute

$$\xi(n_r) = \frac{A}{na} = \frac{R^2}{nr^2} = \frac{[1 + 2\|t_3\|(n_r - 1)]^2}{4[1 + 3(n_r - 1)n_r]}. \tag{3.65}$$

For example, $\xi(n_r)$ is approximately equal to

$$\{0.60, \quad 0.68, \quad 0.75, \quad 0.80\},$$

when $n_r \to \{2, 3, 6, \infty\}$. Note that ξ grows with the number of rings n_r. The substitution of ξ on μ_n produces a complex expression, which we shall not attempt to differentiate. Instead, the next example gives an idea on the dependence of μ_n on n_r.

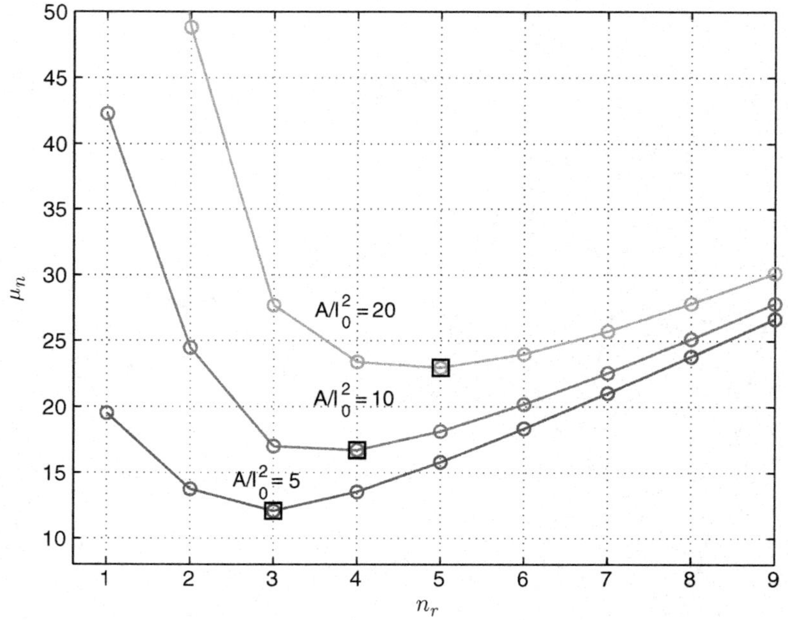

Figure 3.42: Three-bar tensegrity plate: μ_n versus n_r, for $\epsilon = 0.05$ and $A/\ell_0^2 = \{5, 10, 20\}$, from (3.62) using (3.64) and (3.65)

Example 3.18

The formula (3.62) gives the total mass as a function of n. For hexagonal flat plates we plot μ_n versus the number of rings n_r in Figure 3.42 with $\epsilon = 0.05$ and various aspect ratio parameters $A/\ell_0^2 = \{5, 10, 20\}$. The relationship between n and n_r was computed from (3.64) and the factor $\xi(n_r)$ from (3.65). As before, the minimal mass is indicated by squares.

3.9 Non-minimal Regular Prisms

The prism discussed in the previous section has p bars and $3p$ strings and a pretty rigid set of equilibrium conditions. In particular, a unique twist angle α is possible for each p. If one is willing to add strings, then more flexible equilibrium conditions can be established, with possible mechanical advantages. For instance, in this section we discuss adding *diagonal* strings. In Figure 3.29(a), one diagonal string is connected to nodes $\mathbf{n}_5 - \mathbf{n}_1$, and in Figure 3.29(b), one diagonal string is connected to nodes $\mathbf{n}_9 - \mathbf{n}_1$. This produces the prism depicted in Figure 3.43 (also Figure 1.22).

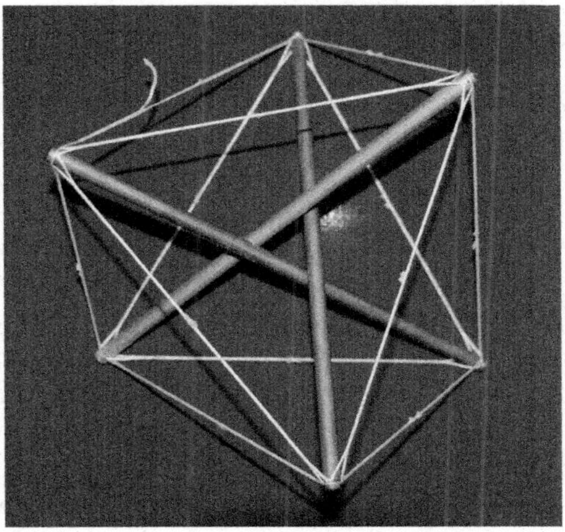

Figure 3.43: A regular non-minimal tensegrity prism. This structure has 12 strings, including 3 extra diagonal strings

3.9.1 Equilibrium

Let γ_d be the force density in the diagonal strings, then

$$\begin{pmatrix} \gamma_t \\ \gamma_b \\ \gamma_v \\ \gamma_d \end{pmatrix} = \frac{\lambda_b}{\cos(\alpha - \pi/p)} \begin{pmatrix} \rho^{-1}\cos(\pi/p) \\ \rho\cos(\pi/p) \\ 2\cos(\alpha)\cos(\pi/p) \\ -\cos(\alpha + \pi/p) \end{pmatrix}, \tag{3.66}$$

where now the twist angle is any angle between

$$\frac{\pi}{2} - \frac{\pi}{p} \leq \alpha \leq \frac{\pi}{2}. \tag{3.67}$$

In this range, one can verify that the γ's are all nonnegative for $\lambda_b > 0$. As in the minimal regular prism, $\gamma_b/\gamma_t = \rho^2$.

Note also that for $\alpha = \pi/2 - \pi/p$, $\gamma_d = 0$. That is, diagonal strings are not needed when the canonical twist angle of a minimal prism is chosen. Conversely, vertical strings are not needed when $\alpha = \pi/2$. Note that going all the way to $\pi/2$ may cause the bars to collide, and hence may not be realizable.

From the above equilibrium conditions, and using the methods of Section 2.5, one can verify that the above prism is stiff in the sense that no finite or infinitesimal mechanisms are present for any twist angle inside the feasible interval (3.67). Indeed, columns built using the self-similar principle discussed in (3.7) using such stiff non-minimal tensegrity prisms as basic units will also be free of infinitesimal mechanisms.

3.10 Chapter Summary

In this chapter we have introduced the concept of complexity in the context of self-similar iterations. Since the number of self-similar iterations fixes the number of discrete structural elements (bars and tendons), we refer to the number of self-similar iterations as the *complexity* of the system. We have shown that the self-similar structure that minimizes mass under compressive loads is obtained from the *T-Bar* self-similar rule. Furthermore, the optimal complexity of such a structure under compression is finite (as opposed to the optimal bending result of Michell which gives infinite complexity. See Chapter 4). In the case of compressive structures, the optimal complexity, p, is the smaller of the two numbers p computed from (3.26) or (3.21). One of these numbers, (3.21), is independent of external force or material properties, while the other, (3.26), depends on both. Thus when failure is by buckling then the optimal topology of compressive members remains independent of the material and external load, whereas when failure of compressive members is by material yielding, then the optimal topology depends on material and external load.

As in the case of the *T-Bar* topology, the optimal compressive structure using the *D-Bar* self-similar rule also reduces mass, up to a finite number of iterations. As in the *T-Bar* case, explicit formulas are given for the optimal number of self-similar iterations (3.43). Since the geometric arrangements of the compressive and tensile elements are different in the *T-Bar* and *D-Bar* cases, we have shown how to use the advantages of both to build a structure that collapses and deploys easily with simple tendon control. If one wishes to design a deployable structure that maintains the same cross-sectional shape of the original structure, but expands its length like an accordion, one can use the *T-Bar* self-similar rule except for the very last iteration, which would be a single *D-Bar* iteration.

Tensegrity prisms were discussed and equilibrium formulas have been derived for minimal and non-minimal regular prisms. By combining tensegrity prisms we analyzed two types of self-similar structures: class 2 tensegrity columns and class 1 plates. Class 2 tensegrity columns are a natural generalization of the planar unit-self-similar design using *D-Bar* or *T-Bar* units to the three-dimensional case. Formulas for the design of plates were obtained in the case of distributed compressive crushing loads. Class 1 plates were designed so that the connection between units happened at an existing straight string. This way of connecting preserves the equilibrium of the individual units but also may come at the expense of low stiffness. Indeed, if a unit is designed with zero pretension, then the stiffness between units will be zero. One can overcome this problem by allowing the units to overlap, thus avoiding the connection at straight strings. This construction will be reported in another publication.

It is interesting to observe that, in the presence of a crushing force, the minimal mass of the two types of plates, topologies A and B, is the same, and the optimal complexity (density of units filling the space) is the same. Topology A has a more uniform distribution of the nodes than topology B, and the stiffness of the two plates might be different, but stiffness was not computed in these examples.

3.11 Advanced Material

3.11.1 Equilibrium of Regular p-Bar Tensegrity Prism

The notation in this section refers to Figure 3.29. We have that p is the number of bars per prism, h is the prism height, r_t and r_b are the top and bottom radii, respectively, α is the twist angle, and $\phi = \text{sign}(\alpha)2\pi/p$ is the prism characteristic angle (see Section 3.6.1 for more details).

Define the rotation matrix

$$R(\phi) := \begin{bmatrix} \cos(\phi) & -\sin(\phi) & 0 \\ \sin(\phi) & \cos(\phi) & 0 \\ 0 & 0 & 1 \end{bmatrix},$$

then all *bottom nodes* of a prism can be located by

$$\mathbf{n}_i = R(\phi)^i \begin{pmatrix} r_b \\ 0 \\ 0 \end{pmatrix}, \qquad i = 0, \ldots, p-1,$$

whereas the *top nodes* are

$$\mathbf{n}_i = R(\phi)^{i-p+1} \begin{pmatrix} r_t \cos(\alpha) \\ -r_t \sin(\alpha) \\ h \end{pmatrix}, \qquad i = p, \ldots, 2p-1.$$

Using the above nodes all members can be precisely located. Namely, bars are given by

$$\mathbf{b}_i = \mathbf{n}_{p+i} - \mathbf{n}_i, \qquad 0 \le i \le p-1,$$

bottom strings:

$$\mathbf{s}_{bi} = \mathbf{n}_{i+1 \pmod p} - \mathbf{n}_i, \qquad 0 \le i \le p-1,$$

top strings:

$$\mathbf{s}_{ti} = \mathbf{n}_{p+[i+1 \pmod p]} - \mathbf{n}_{p+i}, \qquad 0 \le i \le p-1,$$

vertical strings:

$$\mathbf{s}_{vi} = \mathbf{n}_{p+[i-1 \pmod p]} - \mathbf{n}_i, \qquad 0 \le i \le p-1,$$

and diagonal strings:

$$\mathbf{s}_{di} = \mathbf{n}_{p+[i-2 \pmod p)]} - \mathbf{n}_i, \qquad 0 \le i \le p-1.$$

For equilibrium (see Section 2.4 for details), the sum of the forces at a bottom node i must be zero, i.e.,

$$\lambda_b \mathbf{b}_i = \gamma_b[\mathbf{s}_{bi} - \mathbf{s}_{b(i-1)}] + \gamma_v \mathbf{s}_{vi} + \gamma_d \mathbf{s}_{di}.$$

Because of symmetry we can consider a common force density for each group of strings instead of individual force coefficients. Symmetry will also allow us to analyze equilibrium at a single bottom node and single top node. At the ith bottom node the above equilibrium equation can be written as

$$\mathbf{A}_{bi}\mathbf{x} = \lambda_b \mathbf{b}_i, \qquad \mathbf{A}_{bi} := \begin{bmatrix} \mathbf{s}_{bi} - \mathbf{s}_{b(i-1)} & \mathbf{s}_{vi} & \mathbf{s}_{di} \end{bmatrix}, \qquad \mathbf{x} := \begin{bmatrix} \gamma_b \\ \gamma_v \\ \gamma_d \end{bmatrix}.$$

Therefore,

$$\mathbf{x} = \lambda_b \mathbf{A}_{bi}^{-1} \mathbf{b}_i = \frac{\lambda_b}{\cos(\alpha - \pi/p)} \begin{pmatrix} (r_t/r_b)\cos(\pi/p) \\ 2\cos(\alpha)\cos(\pi/p) \\ -\cos(\alpha + \pi/p) \end{pmatrix}.$$

The sum of the forces at a top node $p+i$ is

$$-\lambda_b \mathbf{b}_i = \gamma_t[\mathbf{s}_{ti} - \mathbf{s}_{t(i-1)}] - \gamma_v \mathbf{s}_{v(i+1)} - \gamma_d \mathbf{s}_{d(i+2)},$$

which can be written as

$$\mathbf{A}_{ti}\mathbf{y} = -\lambda_b \mathbf{b}_i, \quad \mathbf{A}_{bi} := \begin{bmatrix} \mathbf{s}_{ti} - \mathbf{s}_{t(i-1)} & -\mathbf{s}_{v(i+1)} & -\mathbf{s}_{d(i+2)} \end{bmatrix}, \quad \mathbf{y} := \begin{bmatrix} \gamma_t \\ \gamma_v \\ \gamma_d \end{bmatrix}.$$

Therefore,

$$\mathbf{y} = -\lambda_b \mathbf{A}_{ti}^{-1} \mathbf{b}_i = \frac{\lambda_b}{\cos(\alpha - \pi/p)} \begin{pmatrix} (r_b/r_t)\cos(\pi/p) \\ 2\cos(\alpha)\cos(\pi/p) \\ -\cos(\alpha + \pi/p) \end{pmatrix}.$$

Note that γ_v and γ_d appear in both $\mathbf{x}$ and $\mathbf{y}$ but, as expected, with the same value. For stable equilibrium (see Section 2.4), we must also have

$$\mathbf{x} \ge 0, \qquad\qquad \mathbf{y} \ge 0.$$

Since $\cos(\pi/p) \ge 0$ for all $p \ge 2$, we must have

$$\lambda_b \ge 0, \qquad \cos(\alpha - \pi/p) \ge 0 \quad \Rightarrow \quad \pi/p - \pi/2 \le \alpha \le \pi/p + \pi/2$$

for positivity of γ_b and γ_t. For positivity of γ_v and γ_d, we must have

$$\begin{aligned} \cos(\alpha) \geq 0 \quad &\Rightarrow \quad -\pi/2 \leq \alpha \leq \pi/2, \\ \cos(\alpha + \pi/p) \leq 0 \quad &\Rightarrow \quad \pi/2 - \pi/p \leq \alpha \leq 3\pi/2 - \pi/p. \end{aligned}$$

That is,

$$\pi/2 - \pi/p \leq \alpha \leq \pi/2.$$

When $\alpha = \pi/2 - \pi/p$, we have $\gamma_d = 0$; and when $\alpha = \pi/2$, we have $\gamma_v = 0$.

3.11.2 Tensegrity Plates

Topology A

Subtracting (3.57) from (3.56) we obtain the equation

$$\mathbf{n}_0 - \mathbf{n}_5 = \beta_b \mathbf{n}_2 + (1 - \beta_b)\mathbf{n}_1 - \beta_t \mathbf{n}_3 - (1 - \beta_t)\mathbf{n}_4,$$

which is independent of $\mathbf{t}_{3A}$ and involves only nodes of the first unit. This equation can be rewritten as the linear algebra problem

$$\mathbf{A}_p \mathbf{x} = \mathbf{b}_p,$$

where

$$\mathbf{A}_p := \begin{bmatrix} \mathbf{n}_2 - \mathbf{n}_1 & \mathbf{n}_4 - \mathbf{n}_3 \end{bmatrix}, \quad \mathbf{x} := \begin{bmatrix} \beta_b \\ \beta_t \end{bmatrix}, \quad \mathbf{b}_p := \mathbf{n}_0 - \mathbf{n}_1 - \mathbf{n}_5 + \mathbf{n}_4.$$

The solution to this problem is

$$\begin{bmatrix} \beta_b \\ \beta_t \end{bmatrix} = \mathbf{x} = \mathbf{A}_p^{-1}\mathbf{b}_p = \frac{1}{2}\left(1 + \sqrt{3}\tan(\alpha/2)\right)\begin{bmatrix} 1 \\ 1 \end{bmatrix},$$

from which, using (3.56), we obtain

$$\mathbf{t}_{3A} = \beta_b \mathbf{n}_2 + (1 - \beta_b)\mathbf{n}_1 - \mathbf{n}_0 = -\frac{3r}{2}\begin{bmatrix} 1 \\ \tan(\alpha/2) \end{bmatrix}.$$

We obtain (3.58) after substituting $\alpha = \pi/6$, since the prisms are minimal and regular.

Topology B

Summing (3.56) and (3.60), we obtain the equation

$$\mathbf{n}_0 + \mathbf{n}_3 = \beta_b \mathbf{n}_2 + (1 - \beta_b)\mathbf{n}_1 + \beta_t \mathbf{n}_4 + (1 - \beta_t)\mathbf{n}_5,$$

which is independent of $\mathbf{t}_{3B}$ and involves only nodes of the first unit. This equation can be rewritten as the linear algebra problem

$$\mathbf{A}_p \mathbf{x} = \mathbf{b}_p,$$

where

$$\mathbf{A}_p := \begin{bmatrix} \mathbf{n}_2 - \mathbf{n}_1 & \mathbf{n}_4 - \mathbf{n}_5 \end{bmatrix}, \qquad \mathbf{x} := \begin{bmatrix} \beta_b \\ \beta_t \end{bmatrix}, \qquad \mathbf{b}_p := \mathbf{n}_0 + \mathbf{n}_3 - \mathbf{n}_1 - \mathbf{n}_5.$$

The solution to this problem is

$$\begin{bmatrix} \beta_b \\ \beta_t \end{bmatrix} = \mathbf{x} = \mathbf{A}_p^{-1} \mathbf{b}_p = \left(\frac{3 - 6\cos(\alpha)}{\sqrt{3}\sin(\alpha) - 3\cos(\alpha)} - 1 \right) \begin{bmatrix} 1 \\ 1 \end{bmatrix},$$

from which, using (3.56), we obtain

$$\mathbf{t}_{3B} = \beta_b \mathbf{n}_2 + (1 - \beta_b)\mathbf{n}_1 - \mathbf{n}_0 = \frac{3}{2} \begin{pmatrix} -1 \\ \dfrac{\sqrt{3}\cos(\alpha) + 3\sin(\alpha) - 2\sqrt{3}}{\sqrt{3}\sin(\alpha) - 3\cos(\alpha)} \end{pmatrix}.$$

We obtain (3.61) after substituting $\alpha = \pi/6$, since the prisms are minimal and regular.

Chapter 4

Design of Bending Structures

This chapter provides the complete analytical solution for the design of a planar cantilevered structure to support a given load, yielding the smallest volume of material required to support the load. Constraints against both types of material failure, yielding and buckling, are guaranteed.

The most important reference in this chapter is the seminal work of Michell in 1904 [Mic04], where he showed a continuum of material to minimize volume for a material system under bending loads, under the assumption that all members use the same materials. (Michell's optimization result requires that all members are composed of the same material, as first pointed out by Rozvany [Roz96, Roz97].) This is commonly called today the *Michell Truss* [HP69, Sch81, OS01, Roz98]. This chapter gives a discrete optimal result for beams in bending, whereas Michell's 1904 results [Mic04] provided only the continuum solution.

4.1 Michell Topology

We start with some definitions of spirals and then show how to connect them together to make an interesting topology. We shall use Michell's name to label the components of our technique (*Michell Spirals* and *Michell Topology*), even though such concepts are new.

4.1.1 Michell Spirals

Consider a sequence of lines of length $p_\ell, p_{\ell+1}, \ldots$ connected end to end as shown in Figure 4.1. Relative to a common origin, $\mathbf{0}$, the geometry of these connections can be described as follows.

R.E. Skelton, M.C. de Oliveira, *Tensegrity Systems*,
DOI 10.1007/978-0-387-74242-7_4,

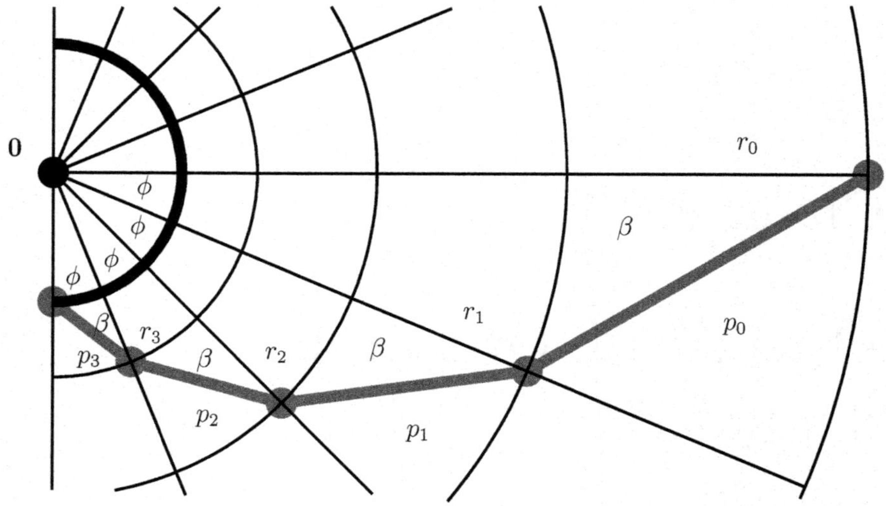

Figure 4.1: A Michell Spiral of order 4 ($\phi = \pi/16$, $\beta = \pi/6$)

Definition 4.1 *Let r_ℓ define a set of radii from a common origin, $\mathbf{0}$, for $\ell = 0, 1, 2, \ldots, q$. Let $p_\ell, \ell = 0, 1, 2, \ldots, q-1$, define the lengths of lines beginning at points with radius r_ℓ and terminating at points with radius $r_{\ell+1}$. Then a* Michell Spiral *of order q is defined by the end-to-end connections of lines of length p_ℓ, satisfying,*

$$r_{\ell+1} = a r_\ell, \qquad p_\ell = c r_\ell, \qquad \ell = 0, 1, 2, \ldots, q, \tag{4.1}$$

where $a > 0$ and $c > 0$.

If

$$a = \frac{\sin\beta}{\sin(\beta+\phi)}, \qquad c = \frac{\sin\phi}{\sin(\beta+\phi)}, \tag{4.2}$$

then the sequence generates a Michell Spiral as in Figure 4.1. The relations between (a, c) and (ϕ, β) given above follow from Figure 4.1 by observing that

$$r_{\ell+1}\cos\phi + p_\ell\cos\beta = r_\ell, \tag{4.3}$$

$$r_{\ell+1}\sin\phi = p_\ell\sin\beta. \tag{4.4}$$

Obviously if $a = 1$, all radii have the same value and the lines p_ℓ are secants of a circle. In this case $\phi + 2\beta = \pi$. For the spirals to converge to the origin one needs $a < 1$, which corresponds to $\phi + 2\beta < \pi$.

4.1.2 Michell Topology

Consider Figure 4.2, where a Michell Spiral of order 4 is described by the connection of nodes $\mathbf{n}_{00}$, $\mathbf{n}_{01}$, $\mathbf{n}_{02}$, $\mathbf{n}_{03}$, $\mathbf{n}_{04}$. A Michell Spiral of order 3 is

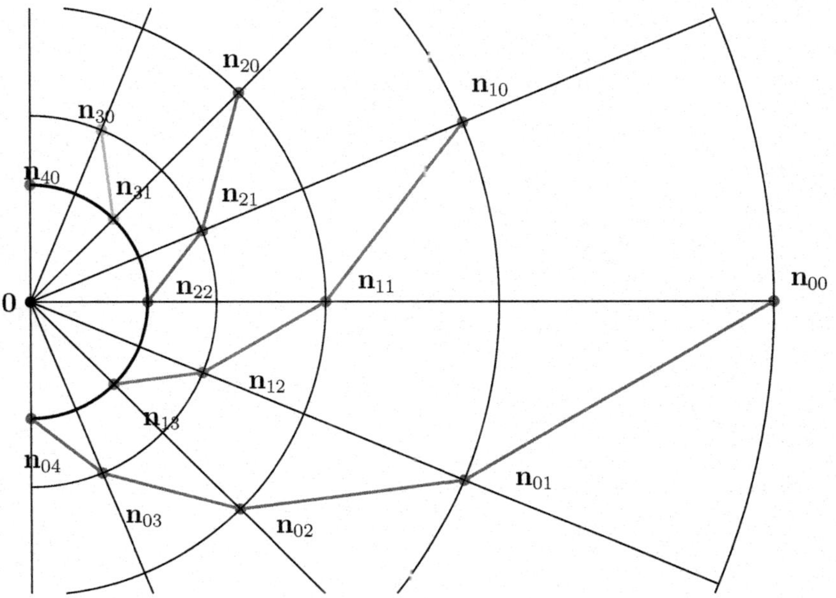

Figure 4.2: Michell Spirals of order 0, 1, 2, 3, and 4 ($\phi = \pi/16$, $\beta = \pi/6$)

described by the connections of nodes $\mathbf{n}_{10}$, $\mathbf{n}_{11}$, $\mathbf{n}_{12}$, $\mathbf{n}_{13}$. A Michell Spiral of order 1 is described by the straight line between nodes $\mathbf{n}_{30}$ and $\mathbf{n}_{31}$, a Michell Spiral of order 0 is the single node $\mathbf{n}_{40}$, and so forth.

Relative to the common origin $\mathbf{0}$, nodes lying on the same radius have the same magnitude, that is,

$$\|\mathbf{n}_{ik}\| = \|\mathbf{n}_{mn}\|, \quad \text{for all } i + k = m + n. \tag{4.5}$$

From Figure 4.2, nodes with the same radius are related by a phase shift of $2m\phi$ where m is an integer. This means that

$$\mathbf{n}_{i+m,k-m} = e^{j2m\phi}\mathbf{n}_{ik}. \tag{4.6}$$

Note that in this chapter we use complex notation (phasor) to describe vectors in a plane, instead of the more general three-dimensional setup of Chapter 2. This will simplify many of the derivations we carry on in this chapter.

The members of the spiral are described by defining the vector connecting nodes $\mathbf{n}_{ik}$ and $\mathbf{n}_{i,k+1}$:

$$\mathbf{m}_{ik} = \mathbf{n}_{ik} - \mathbf{n}_{i,k+1}, \tag{4.7}$$

where the vector $\mathbf{n}_{ik}$ has magnitude and phase given by

$$\mathbf{n}_{ik} = n_{ik}e^{j\varphi_{n_{ik}}}, \qquad n_{ik} = r_{i+k}, \qquad \varphi_{n_{ik}} = (i-k)\phi, \tag{4.8}$$

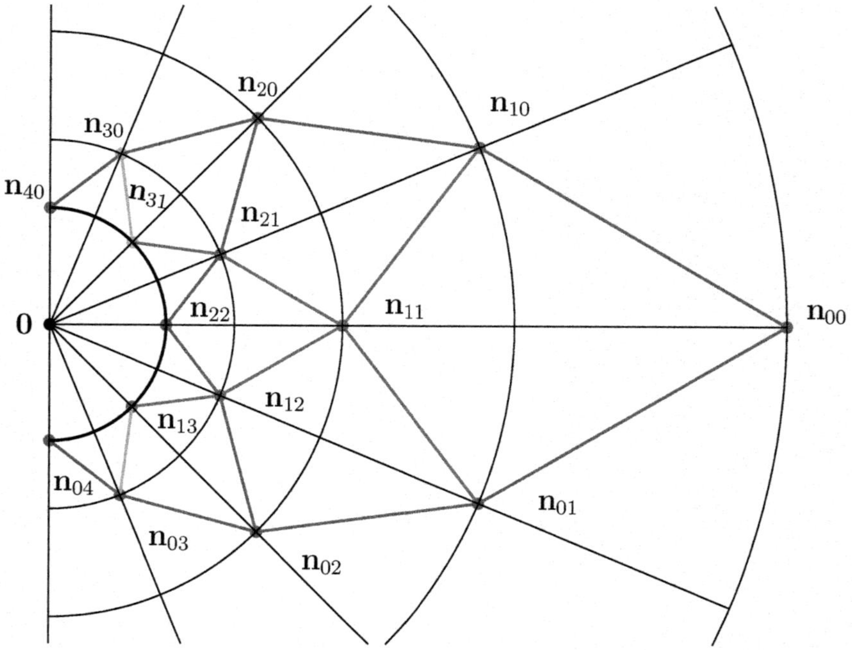

Figure 4.3: Michell Topology of order 4 ($\phi = \pi/16$, $\beta = \pi/6$)

where r_k satisfies (4.1) for some specified r_0.

The mirror image of all lines reflected about the axis $\mathbf{0} - \mathbf{n}_{00}$ is obtained after computing the conjugate of the vectors $\mathbf{m}_{ik}$

$$\overline{\mathbf{m}}_{ik} = \overline{\mathbf{n}_{ik} - \mathbf{n}_{i,k+1}} = \mathbf{n}_{ki} - \mathbf{n}_{k+1,i}. \tag{4.9}$$

Note also that (4.7) and (4.8) yield

$$\mathbf{m}_{ik} = p_{i+k} e^{j[\beta+(i-k)\phi]}, \tag{4.10}$$

where p_m ($m = 0, 1, 2, \ldots, q$) satisfies (4.1).

The collection of members $\mathbf{m}_{ik}$ and their conjugates form the final structure of the Michell Topology of order 4 as shown in Figure 4.3.

Definition 4.2 *A* Michell Topology *of order q is described by the Michell Spirals of order $\leq q$ and their conjugate spirals where* (4.5) *and* (4.6) *hold (Figure* 4.3 *illustrates for $q = 4$).*

Note the importance on the parameters (ϕ, β) or, equivalently, (a, c) in the shape of the Michell Topology. Figure 4.4 illustrates two different choices of parameters for $q = 8$. Note that as the product $q\phi \to 0$ the Michell Topology approaches a standard triangular-shaped trellis topology.

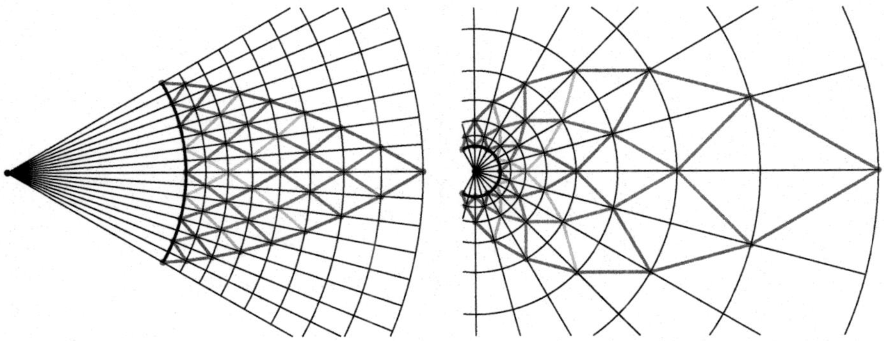

(a) ($\phi=\pi/48, \beta=\pi/6$) (b) ($\phi=\pi/12, \beta=\pi/6$)

Figure 4.4: Michell Topologies of order 8

4.2 Michell Topology in Static Equilibrium

4.2.1 Force Equilibrium at a Generic Node

Following the development in the previous section, we shall write the force vector $\mathbf{w}_{ik}$ applied to node $\mathbf{n}_{ik}$ in complex form

$$\mathbf{w}_{ik} = w_{ik} e^{j\varphi_{w_{ik}}}. \tag{4.11}$$

We choose to describe the angle $\varphi_{w_{ik}}$ in terms of an angle θ_{ik}, measured relative to the radial line to node $\mathbf{n}_{ik}$, that is,

$$\varphi_{w_{ik}} = \theta_{ik} + (i-k)\phi, \tag{4.12}$$

where θ_{ik} is the angle at which the external force $\mathbf{w}_{ik}$ is applied, measured from the radial line to node $\mathbf{n}_{ik}$.

As in Section 2.4, at node $\mathbf{n}_{ik}$ (illustrated in Figure 4.5), we shall sum force vectors entering the node, where the forces along directions $\mathbf{m}_{ik}$, $\overline{\mathbf{m}}_{ki}$, $\mathbf{m}_{i,k-1}$, $\overline{\mathbf{m}}_{k,i-1}$, respectively, have magnitudes, f_{ik}, t_{ki}, $f_{i,k-1}$, $t_{k,i-1}$. Using these, the sum of forces at node $\mathbf{n}_{ik}$ yields,

$$f_{ik}\frac{\mathbf{m}_{ik}}{\|\mathbf{m}_{ik}\|} + t_{ki}\frac{\overline{\mathbf{m}}_{ki}}{\|\mathbf{m}_{ki}\|} - f_{i,k-1}\frac{\mathbf{m}_{i,k-1}}{\|\mathbf{m}_{i,k-1}\|} - t_{k,i-1}\frac{\overline{\mathbf{m}}_{k,i-1}}{\|\mathbf{m}_{k,i-1}\|} + \mathbf{w}_{ik} = 0. \tag{4.13}$$

As shown at the end of the chapter, for equilibrium one should solve the recursive equations

$$\begin{pmatrix} t_{ki} \\ f_{ik} \end{pmatrix} p_{i+k} = \mathbf{\Omega} \begin{pmatrix} t_{k,i-1} \\ f_{i,k-1} \end{pmatrix} p_{i+k-1} + \mathbf{\Phi}_{ik} w_{ik}, \tag{4.14}$$

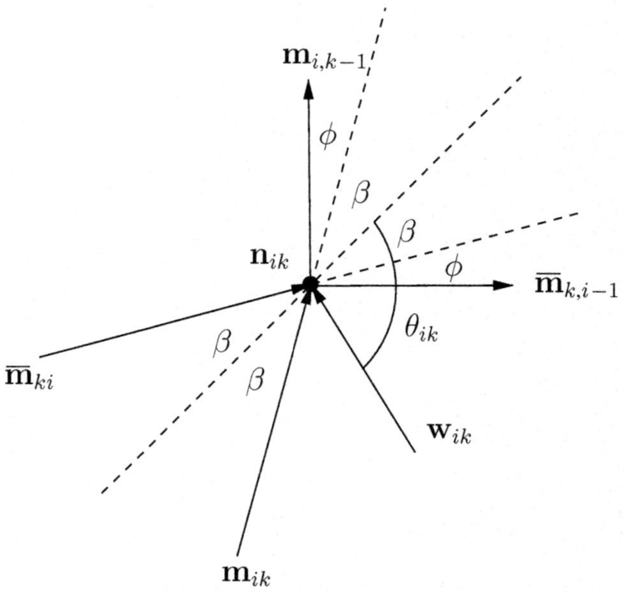

Figure 4.5: Sum of the forces at node $\mathbf{n}_{ik}$

with "initial" conditions

$$\begin{pmatrix} t_{00} \\ f_{00} \end{pmatrix} p_0 = \mathbf{\Phi}_{00} w_{00}, \tag{4.15}$$

where,

$$\mathbf{\Phi}_{ik} = \frac{p_{i+k}}{\sin(2\beta)} \begin{bmatrix} \sin(\theta_{ik} - \beta) \\ -\sin(\theta_{ik} + \beta) \end{bmatrix}, \tag{4.16}$$

$$\mathbf{\Omega} = \frac{1}{2} \begin{bmatrix} g & -h \\ -h & g \end{bmatrix}, \qquad g = 1 + \frac{\tan\beta}{\tan(\beta + \phi)}, \quad h = \frac{\sin\phi}{\cos\beta \sin(\beta + \phi)}. \tag{4.17}$$

It is straightforward to show that $g + h = 2$, which will be useful later.

Example 4.1

Let $q = 1$, in which case $\mathbf{n}_{00}$ is the only node at which external forces are applied

$$\begin{pmatrix} t_{00} \\ f_{00} \end{pmatrix} = \frac{w_{00}}{\sin(2\beta)} \begin{bmatrix} \sin(\theta_{00} - \beta) \\ -\sin(\theta_{00} + \beta) \end{bmatrix}.$$

Note that the forces do not depend directly on p_0 or r_0. The dependence on the length of the member has been transferred to the angle β.

In the case $0 \leq \beta < \pi/2$ three cases are of interest. The first is when $|\theta_{00}| < \beta < \pi/2$, then $\sin(2\beta) \geq 0$ and

$$\begin{aligned} -\pi < -2\beta < \theta_{00} - \beta < 0 \quad &\Longrightarrow \quad \sin(\theta_{00} - \beta) < 0 \\ 0 < \theta_{00} + \beta < 2\beta < \pi \quad &\Longrightarrow \quad \sin(\theta_{00} + \beta) > 0, \end{aligned}$$

which implies $t_{00} < 0$ and $f_{00} < 0$. That is, both members of the truss are in *tension*.

The second is when $|\theta_{00} - \pi| < \beta < \pi/2$. A similar analysis shows that $t_{00} > 0$ and $f_{00} > 0$. That is, both members of the truss are in *compression*.

Finally, when $\beta < |\theta_{00}| < \pi - \beta$ then $t_{00} f_{00} < 0$, indicating that one member of the truss is in compression while the other is in tension. In this case the truss is said to be in *bending*.

The scenario is significantly more complicated when $q > 1$ and in the presence of forces in any nodes of the Michell Topology. Yet, as the next section will show, it is possible to derive a remarkably simple recursive formula that allows one to compute the forces in all nodes of the topology given any set of external forces.

4.2.2 Linear Propagation of Forces

Define a vector $\mathbf{x}_\alpha \in \mathbb{R}^{2(\alpha+1)}$ which contains forces (normalized by multiplying by the length of the member) in all members that lie within the radii r_α and $r_{\alpha+1}$. That is,

$$\mathbf{x}_\alpha = \begin{bmatrix} t_{0\alpha} \\ f_{\alpha 0} \\ t_{1,\alpha-1} \\ f_{\alpha-1,1} \\ \vdots \\ t_{i,\alpha-i} \\ f_{\alpha-i,i} \\ \vdots \\ t_{\alpha,0} \\ f_{0,\alpha} \end{bmatrix} p_\alpha. \tag{4.18}$$

For example, the normalized forces in all members between radii r_0 and r_1, between radii r_1 and r_2, between radii r_2 and r_3, and between radii r_3 and

r_4, are, respectively,

$$\mathbf{x}_0 = \begin{bmatrix} t_{00} \\ f_{00} \end{bmatrix} p_0, \quad \mathbf{x}_1 = \begin{bmatrix} t_{01} \\ f_{10} \\ t_{10} \\ f_{01} \end{bmatrix} p_1, \quad \mathbf{x}_2 = \begin{bmatrix} t_{02} \\ f_{20} \\ t_{11} \\ f_{11} \\ t_{20} \\ f_{02} \end{bmatrix} p_2, \quad \mathbf{x}_3 = \begin{bmatrix} t_{03} \\ f_{30} \\ t_{12} \\ f_{21} \\ t_{21} \\ f_{12} \\ t_{30} \\ f_{03} \end{bmatrix} p_3. \tag{4.19}$$

Using the recursive relations (4.14–4.17) it is straightforward to show that the vectors $\mathbf{x}_\alpha$ and $\mathbf{x}_{\alpha+1}$ are related by the recursive form,

$$\mathbf{x}_{\alpha+1} = \mathbf{A}_\alpha \mathbf{x}_\alpha + \mathbf{B}_\alpha \mathbf{u}_\alpha, \qquad \alpha = 0, 1, 2, \ldots, q-1, \tag{4.20}$$

where

$$\mathbf{A}_\alpha \in \mathbb{R}^{2(\alpha+2)\times 2(\alpha+1)}, \quad \mathbf{B}_\alpha \in \mathbb{R}^{2(\alpha+2)\times(\alpha+2)}, \quad \mathbf{x}_\alpha \in \mathbb{R}^{2(\alpha+1)}, \quad \mathbf{u}_\alpha \in \mathbb{R}^{\alpha+2}.$$

It follows that

$$\mathbf{u}_\alpha = \begin{bmatrix} w_{\alpha+1,0} \\ w_{\alpha,1} \\ w_{\alpha-1,2} \\ w_{\alpha-2,3} \\ w_{\alpha-3,4} \\ \vdots \\ w_{0,\alpha+1} \end{bmatrix} \tag{4.21}$$

and

$$\mathbf{A}_\alpha = \begin{bmatrix} \mathbf{J}_2 & 0 & 0 & 0 & 0 \\ 0 & \mathbf{J} & 0 & 0 & 0 \\ 0 & 0 & \ddots & 0 & 0 \\ 0 & 0 & 0 & \mathbf{J} & 0 \\ 0 & 0 & 0 & 0 & \mathbf{J}_1 \end{bmatrix},$$

$$\mathbf{B}_\alpha = \begin{bmatrix} \mathbf{\Phi}_{\alpha+1,0} & 0 & 0 & 0 & 0 & 0 \\ 0 & \mathbf{\Phi}_{\alpha,1} & 0 & 0 & 0 & 0 \\ 0 & 0 & \mathbf{\Phi}_{\alpha-1,2} & 0 & 0 & 0 \\ 0 & 0 & 0 & \mathbf{\Phi}_{\alpha-2,3} & 0 & 0 \\ 0 & 0 & 0 & 0 & \ddots & 0 \\ 0 & 0 & 0 & 0 & 0 & \mathbf{\Phi}_{0,\alpha+1} \end{bmatrix}, \tag{4.22}$$

with

$$\mathbf{J} = \mathbf{\Omega} \begin{bmatrix} 0 & 1 \\ 1 & 0 \end{bmatrix} = \begin{bmatrix} \mathbf{J}_1 & \mathbf{J}_2 \end{bmatrix}. \tag{4.23}$$

Note that all elements in $\mathbf{B}_\alpha$ have arguments in $\mathbf{\Phi}_{ik}$ such that $i+k=\alpha+1$, and all elements in $\mathbf{u}_\alpha$ have arguments w_{ik} such that $i+k=\alpha+1$.

This proves the following theorem.

Theorem 4.1 *Let a truss be arranged according to the Michell Topology of order q (as in Figure 4.3 for $q=4$) satisfying (4.1), having external forces $\mathbf{w}_{ik}$ applied at the nodes $\mathbf{n}_{ik}$ with $i \geq 0$, $j \geq 0$, and $i+j \leq q$. Let $\mathbf{x}_\alpha$ contain the forces (normalized by the member length) in all members within the band of members between radii r_α and $r_{\alpha+1}$ as in (4.18) and $\mathbf{u}_\alpha$ contain the magnitude of the forces at the nodes with radius r_α as in (4.21). Then the forces propagate from one band to the next according to the linear recursive equation (4.20).*

4.3 Michell Topologies Under a Single Bending Load

As we discussed in the previous section, in general, it is not possible to characterize the direction of the forces in a Michell Topology a priori, as we have done in Example 4.1. In this section we discuss one very important exception: the case of a single load applied at the node $\mathbf{n}_{00}$.

More precisely, we consider a single nonzero external force $\mathbf{w}_{00}=\mathbf{w}=we^{j\theta}$ applied at node $\mathbf{n}_{00}$ of a Michell Topology. We consider only trusses for which $a<1$, or equivalently,

$$\phi > 0, \qquad \beta > 0, \qquad \phi + 2\beta < \pi. \tag{4.24}$$

Note that the above relations imply $\beta < \pi/2$. The main result is the next theorem, proved at the end of the chapter.

Theorem 4.2 *Let a truss be arranged according to the Michell Topology of order q (as in Figure 4.3 for $q=4$) satisfying (4.1) with a and c given by (4.2). Assume the only force applied on the truss is $\mathbf{w}=we^{j\theta}$ at node $\mathbf{n}_{00}$. If*

$$\phi > 0, \qquad 0 < \beta < |\theta| < \pi - \beta, \qquad \phi + 2\beta < \pi,$$

then the forces at all nodes satisfy

$$t_{ik} f_{ik} < 0, \quad \text{for all} \quad i \leq q, \quad k \leq i.$$

Furthermore, if

$$\beta < \theta < \pi - \beta,$$

then $t_{ik} > 0$ for all $i \leq q$, $k \leq i$. Otherwise, if

$$\beta - \pi < \theta < -\beta,$$

then $t_{ik} < 0$ for all $i \leq q$, $k \leq i$.

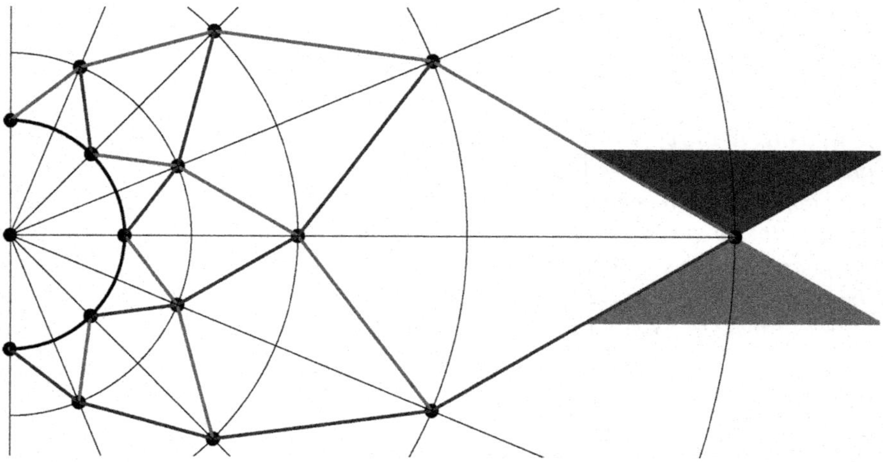

Figure 4.6: Michell Topology of order 4 ($\phi = \pi/16$, $\beta = \pi/6$) showing direction of external force that creates bending; *blue* and *red* indicate a member in compression or tension

The above theorem implies that if the single *external* force of any magnitude lies in the shaded region in Figure 4.6 then the force direction in all *members* of a Michell Topology of any order is known. This direction is the same as in the case of *pure bending*, where $\theta_{00} = \pi/2$. Of course the magnitude of the force in each member will depend upon the magnitude and direction of the external force, but the direction of those member forces will depend on neither magnitude nor direction of the external force. This ability to characterize large regions in the plane where the member forces are *unidirectional* is key to obtaining the optimization results that follow. The remaining sections of this chapter are devoted to such particular classes of loaded Michell Topologies.

4.4 Material Volume of Michell Topologies

In this section we provide an analytical solution to the problem of minimizing the material volume of a Michell Topology with a load in bending, as defined in the previous section, under a yielding failure constraint. We make use of some results previously derived in Section 2.6.1, where we show that minimizing the total material volume V is equivalent to minimizing a certain quantity J, computed as in (2.10), when bars and strings are made of the same material ($\bar{\lambda} = \bar{\gamma}$).

4.4.1 Material Volume for a General Set of External Forces

The results of the previous sections readily apply to general Michell Topologies. Define the following set of indices:

$$\mathcal{B} := \{ik : f_{ik} > 0\}, \qquad \overline{\mathcal{B}} := \{ik : t_{ik} > 0\},$$
$$\mathcal{S} := \{ik : f_{ik} < 0\}, \qquad \overline{\mathcal{S}} := \{ik : t_{ik} < 0\}.$$

The set $\mathcal{B}$ ($\overline{\mathcal{B}}$) represents the set of indices ik of members $\mathbf{m}_{ik}$ ($\overline{\mathbf{m}}_{ik}$) in compression, i.e., all bars. Likewise $\mathcal{S}$ ($\overline{\mathcal{S}}$) represents the set of indices ik of members $\mathbf{m}_{ik}$ ($\overline{\mathbf{m}}_{ik}$) in tension, i.e., strings. Recalling from (4.10) that

$$\|\mathbf{m}_{ik}\| = \|\overline{\mathbf{m}}_{ik}\| = p_{i+k},$$

then

$$\lambda_{ik} = \frac{|t_{ik}|}{\|\mathbf{m}_{ik}\|} = \frac{t_{ik}}{p_{i+k}} \quad \forall\, ik \in \mathcal{B}, \qquad \lambda_{\overline{ik}} = \frac{|f_{ik}|}{\|\overline{\mathbf{m}}_{ik}\|} = \frac{f_{ik}}{p_{i+k}} \quad \forall\, ik \in \overline{\mathcal{B}},$$

and

$$\gamma_{ik} = \frac{|t_{ik}|}{\|\mathbf{m}_{ik}\|} = -\frac{t_{ik}}{p_{i+k}} \quad \forall\, ik \in \mathcal{S}, \qquad \gamma_{\overline{ik}} = \frac{|f_{ik}|}{\|\overline{\mathbf{m}}_{ik}\|} = -\frac{f_{ik}}{p_{i+k}} \quad \forall\, ik \in \overline{\mathcal{S}}.$$

The general formulas of the previous section readily apply after defining matrices $\mathbf{B}$ and $\mathbf{S}$ associated with the sets $\mathcal{B} \cup \overline{\mathcal{B}}$ and $\mathcal{S} \cup \overline{\mathcal{S}}$, respectively. In particular, under the assumption that all bars and strings are made of the same material, one can readily compute the quantity

$$J = (\bar{\lambda} + \bar{\gamma}) \left[\sum_{ik \in \mathcal{B}} t_{ik} p_{i+k} + \sum_{ik \in \overline{\mathcal{B}}} f_{ik} p_{i+k} - \sum_{ik \in \mathcal{S}} t_{ik} p_{i+k} - \sum_{ik \in \overline{\mathcal{S}}} f_{ik} p_{i+k} \right].$$

A case of special interest for this chapter is discussed in the next section.

4.4.2 Michell Topologies Under a Single Bending Load

In this section we discuss the material volume under the single bending load scenario discussed in Section 4.3. For such structure and load configurations, Theorem 4.2 provides means to compute the sign of the forces in the structure for a large range of applied forces, thus determining a priori which members are bars and which members are strings.

For instance, assume that $\beta < \theta < \pi - \beta$. Then from Theorem 4.2 we have that $t_{ik} > 0$ and $f_{ik} < 0$ or, in other words, $\mathcal{B} = \overline{\mathcal{S}} = \emptyset$ and $\overline{\mathcal{B}} = \mathcal{S} = \{ik : \quad 0 \le i \le q, \quad 0 \le k \le i\}$. Similarly, if $\beta - \pi < \theta < -\beta$ then $\mathcal{B} = \overline{\mathcal{S}} = \{ik : \quad 0 \le i \le q, \quad 0 \le k \le i\}$ and $\overline{\mathcal{B}} = \mathcal{S} = \emptyset$.

For the purposes of computing the material volume we can focus on just one scenario, say $\beta < \theta < \pi - \beta$. In this case,

$$\lambda_{ik} = \frac{|t_{ik}|}{\|\mathbf{b}_{ik}\|} = \frac{t_{ik}}{p_{i+k}} \quad \forall\, ik \in \overline{\mathcal{B}}, \qquad \gamma_{ik} = \frac{|f_{ik}|}{\|\mathbf{s}_{ik}\|} = -\frac{f_{ik}}{p_{i+k}} \quad \forall\, ik \in \mathcal{S}.$$

A consequence of the above is that the quantity J can be computed for any Michell Topology of order q simply as

$$J_q = (\bar{\lambda} + \bar{\gamma}) \sum_{i=0}^{q} \sum_{k=0}^{i} (t_{ik} - f_{ik}) p_{i+k}.$$

We shall develop an analytic expression for this formula based on the geometric parameters r_0, ϕ, and β and the external force. We start with the particular case $q = 1$.

Example 4.2

For $q = 1$ and $\beta < \theta < \pi - \beta$ we conclude from the results of Example 4.1 that

$$\begin{aligned} J_1 &= p_0(\bar{\lambda} + \bar{\gamma})(t_{00} - f_{00}), \\ &= r_0\, w_{00}\, (\bar{\lambda} + \bar{\gamma}) \frac{\sin(\theta - \beta) + \sin(\theta + \beta)}{\sin(2\beta)} \frac{\sin\phi}{\sin(\beta + \phi)}, \\ &= r_0\, w_{00}\, (\bar{\lambda} + \bar{\gamma}) \frac{\sin\theta}{\sin\beta} \frac{\sin\phi}{\sin(\beta + \phi)}. \end{aligned}$$

We now return to the general case $q \geq 1$. As shown at the end of the chapter, the fantastically simple relationship holds:

$$J_q = q\, J_1.$$

Following Example 4.2 we have that

$$J_q = q\, r_0\, w_{00}\, (\bar{\lambda} + \bar{\gamma}) \frac{\sin\theta}{\sin\beta} \frac{\sin\phi}{\sin(\beta + \phi)}.$$

If the above analysis is repeated for $\beta - \pi < \theta < -\beta$ then the above formula still holds for θ replacement by $|\theta|$.

The preceding discussion is summarized in the next theorem.

Theorem 4.3 *Let a truss be arranged according to the Michell Topology of order q (as in Figure* 4.3 *for* q = 4*) satisfying* (4.1) *with a and c given by* (4.2)*. Assume the only force applied on the truss is* $\mathbf{w} = w e^{j\theta}$ *at node* $\mathbf{n}_{00}$. *Assume that* $\bar{\lambda} = \bar{\gamma}$ *and*

$$\phi > 0, \qquad 0 < \beta < |\theta| < \pi - \beta, \qquad \phi + 2\beta < \pi.$$

If a Michell Topology of order q has the minimal material volume needed to withstand such a force then the quantity

$$J_q = q\, r_0\, w_{00}\, (\bar{\lambda} + \bar{\gamma}) \frac{\sin|\theta|}{\sin\beta} \frac{\sin\phi}{\sin(\beta+\phi)} \tag{4.25}$$

is also minimal.

4.5 Michell Topologies with Minimum Material Volume Under a Single Bending Load

In this section we will use the analytic formula for J_q obtained in Theorem 4.3 in order to search for parameters that minimize J_q. We start by defining the quantity

$$J'_q := \frac{J_q}{r_0 w_{00}\, (\bar{\lambda} + \bar{\gamma}) \sin|\theta|} = \frac{q\, \sin\phi}{\sin\beta \sin(\beta+\phi)},$$

which we seek to minimize. Note that if the geometric parameters β and ϕ are chosen so as to minimize J'_q, then J_q will simply "scale" the design as a function of the structure's radius (r_0), the external load (w_{00} and $\sin|\theta|$), and the material choice ($\bar{\lambda} = \bar{\gamma}$).

We shall define

$$\rho := \frac{r_q}{r_0} = a^q = \left(\frac{\sin\beta}{\sin(\beta+\phi)}\right)^q$$

and the "complexity" q are specified a priori. The "aspect ratio" of the structure is ρ^{-1}. As shown in the advanced material at the end of the chapter, the optimal choice of ϕ the minimizes J'_q is

$$\phi = \phi^* := \arccos\left(\frac{2}{\rho^{-1/q} + \rho^{1/q}}\right). \tag{4.26}$$

For all $\rho \in [0,1]$ and $q \geq 1$ the value of $2/(\rho^{-1/q} + \rho^{1/q})$ is in the interval $[0,1]$ so that ϕ^* is a well-defined angle in the interval $[0, \pi/2]$. A plot of ϕ^* as a function of ρ for several choices of q is given in Figure 4.7. Note that $\lim_{q\to\infty} \phi^* = 0$.

A more interesting quantity is perhaps the *total angle* $q\,\phi^*$, plotted in Figure 4.8. This angle represents the part of the anchoring disk used to support the structure. For instance, for the truss in Figure 4.3 we have $q\,\phi = \pi/2$ and for the trusses in Figure 4.4 we have $q\,\phi = \pi/6$ and $q\,\phi = 2\pi/3$ for

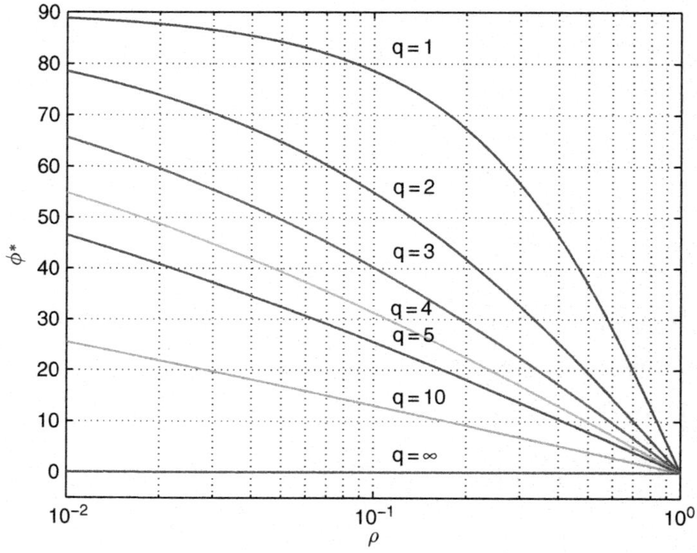

Figure 4.7: The optimal angle ϕ^* (in degrees)

the cases (a) and (b), respectively. Interestingly, even though $\lim_{q\to\infty}\phi^* = 0$, we have that

$$\lim_{q\to\infty} q\,\phi^* = \lim_{q\to\infty} q \arctan\left(\frac{\rho^{-1/q} - \rho^{1/q}}{2}\right) = \ln\rho^{-1}$$

is a finite quantity for $0 < \rho \leq 1$. Therefore, even the infinitely complex Michell Topology built with $q \to \infty$ will have a finite support on the anchoring disk. In particular

$$0.0432 \approx e^{-\pi} \leq \rho \leq 1 \qquad\Rightarrow\qquad \lim_{q\to\infty} q\,\phi^* \leq \pi.$$

That is, the resulting truss has no overlapping members, hence no overlapping material, if the anchoring disk will not be completely encircled by the structure. Note that for any q we will have that $q\,\phi^* \leq \lim_{q\to\infty} q\,\phi^*$, as illustrated in Figure 4.8. Also seen in the figure is that even though it may be possible to construct optimal trusses with no overlapping members by choosing small values of q (in particular optimal trusses with $q = 1$ or $q = 2$ never reach $q\,\phi^* = \pi$ for any choice of $\rho \in (0, 1]$), the total angle quickly approaches π when ρ approaches $e^{-\pi}$. For $q = 10$ this difference is already negligible as compared to $q \to \infty$. In Figure 4.8, $\rho = e^{-\pi}$ is the vertical dotted line.

The exact value of ρ for which $q\,\phi^* = \pi$ as a function of q is computed at the end of the chapter as

$$\rho_\pi := \left(\frac{1}{\cos(\pi/q)} - \tan(\pi/q)\right)^q. \tag{4.27}$$

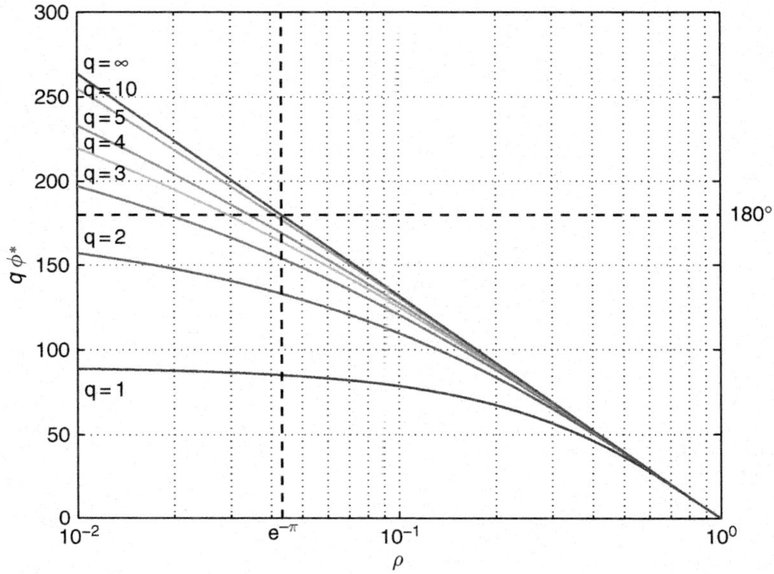

Figure 4.8: The optimal total angle $q\,\phi^*$ (in degrees); material overlap occurs for $\rho \leq e^{-\pi}$

These are the values of ρ at which the curves intersect with the horizontal dotted line $q\,\phi^* = \pi$ in Figure 4.8.

We also show at the end of the chapter that the optimal truss has the invariant property

$$\tan\phi^* \tan(2\beta^*) = 1, \tag{4.28}$$

where

$$\beta^* = \arctan(\rho^{1/q})$$

is the optimal value of β^*. Figure 4.9 shows a plot of β^* as a function of ρ.

With the optimal ϕ^* and β^* we can compute the value of the cost function at the optimum

$$J_q'^* = q\,\rho^{1/q} \sin\phi^* \left(1 + \frac{1}{\tan^2\beta^*}\right) = q\,(\rho^{-1/q} - \rho^{1/q}).$$

This function is plotted in Figure 4.10 as a function of ρ and q.

4.5.1 The Limit as Complexity Grows

In this section we show that the discrete structure that we have considered so far converges to the optimal Michell Truss [Mic04] on the limit when $q \to \infty$.

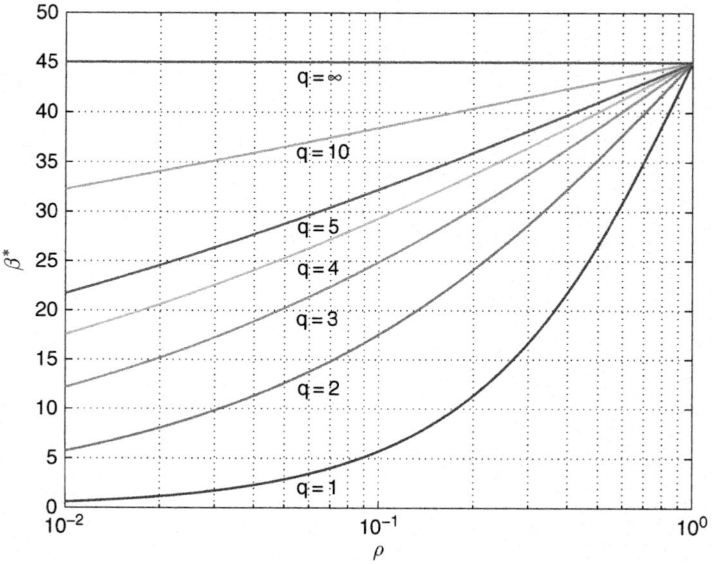

Figure 4.9: The optimal departure angle β^*

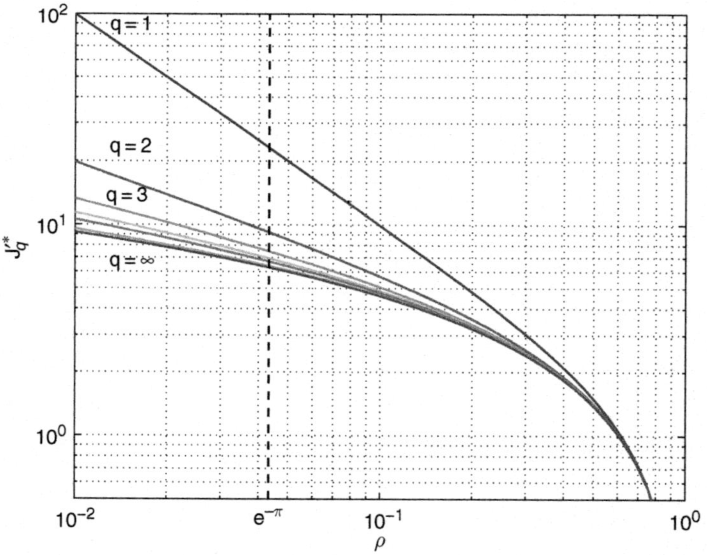

Figure 4.10: The optimal cost ${J'_q}^*$; material overlap occurs for $\rho \leq e^{-\pi}$

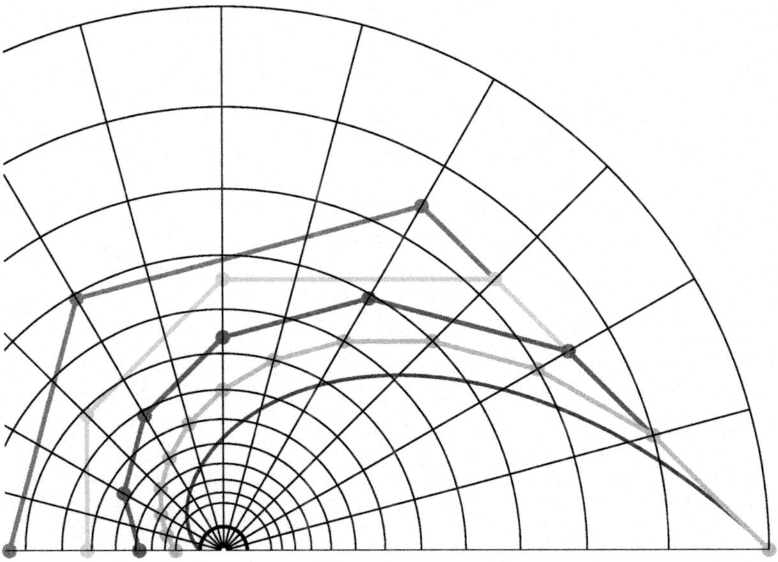

Figure 4.11: Discrete Michell Spirals of order 3, 4, 6, 12, and ∞ (continuous) $(q\,\phi = \pi,\ \beta = \pi/4)$

We start by computing the optimal angle β^*

$$\lim_{q\to\infty} \beta^* = \lim_{q\to\infty} \arctan(\rho^{1/q}) = \frac{\pi}{4}.$$

As we have seen before, spirals on the optimal discrete Michell Truss wrap around the origin only a finite number of times. This is also true even in the case $q \to \infty$ if $\rho > 0$. Indeed

$$\lim_{q\to\infty} q\,\phi^* = \ln \rho^{-1},$$

which is finite. For that to happen we must obviously have $\lim_{q\to\infty} \phi^* = 0$. Thus as $q \to \infty$, mass fills up the region delimited by the continuous spiral shown in Figure 4.11 and its conjugate. The normalized minimum volume ${J'_\infty}^* := \lim_{q\to\infty} {J'_q}^*$ can be computed as

$${J'_\infty}^* = \lim_{q\to\infty} q\,(\rho^{-1/q} - \rho^{1/q}) = 2\,\ln \rho^{-1}.$$

The optimal discrete cost is compared with the limiting case J^*_∞ in Figure 4.12 as a function of ρ and q. Also in Figure 4.11, the continuum spiral is compared with discrete spirals of order 3, 4, and 12. At the end of the chapter we show that the discrete structures we considered so far formally converge to the continuous Michell Truss in the limiting case.

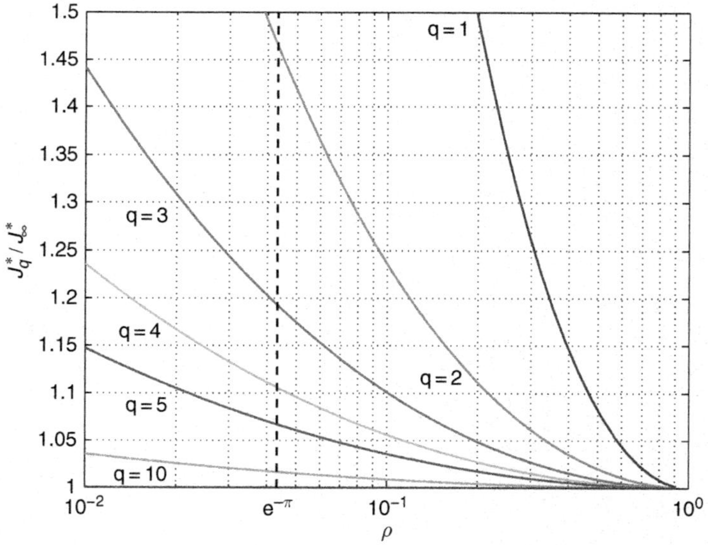

Figure 4.12: The relative optimal discrete cost J_q^*/J_∞^*; material overlap occurs for $\rho \leq e^{-\pi}$

4.5.2 Penalizing Joint Mass Leads to Finite Optimal Complexity

The above theory assumes perfect fabrication with no extra mass added to make the joints. As a practical matter, the joints cannot be made perfectly. Whether glued, or welded, or pinned, the joints will contribute to the mass in proportion to the number of members that must be joined. That is, we add a penalty to the normalized volume criterion J_q' so that the volume of the joints is added to the volume of tensile and compressive material. We assume that the volume of the joints is linearly proportional to the number of members $q(q+1)$, so the new criteria to optimize is

$$J_{\text{total}} = J_q' + \mu\, q(q+1),$$

where μ is the ratio of joint normalized volume to the normalized volume (mass) of the members being joined. Since the extra term depends only on q and not on ϕ, the derivations of the previous sections remain unaltered so that one can compute the normalized minimum total volume as

$$J_{\text{total}}^* = J_q'^* + \mu\, q(q+1) = q\,(\rho^{1/q} - \rho^{-1/q}) + \mu\, q(q+1).$$

In Figure 4.13 we plot this optimal cost for various choices of μ as a function of complexity q for the worst case $\rho = e^{-\pi}$, which represents the limit for which a structure can be constructed with no material overlap for any value of q. Note that a fairly clean joint ($\mu = 0.001$, i.e., the joint mass is 0.1% of

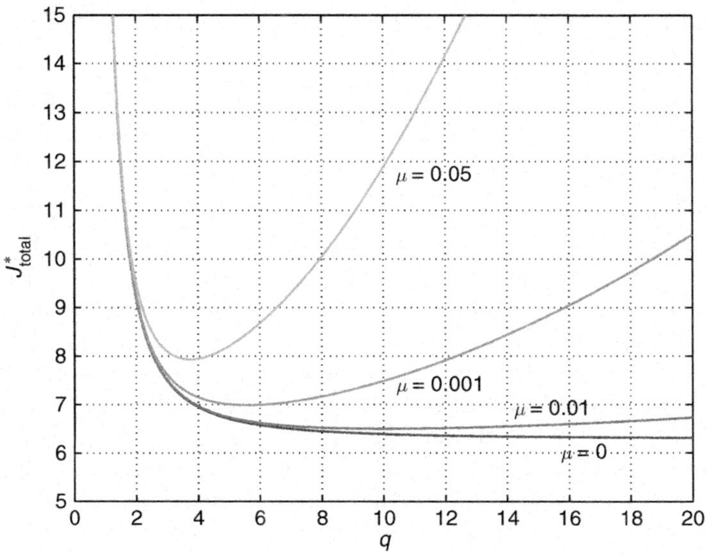

Figure 4.13: The joint mass optimal discrete cost J^*_{total} plotted for the worst case $\rho = e^{-\pi}$

the mass of the member being joined) yields an optimal complexity at around $q = 10$. Note also that more massive joints ($\mu = 0.01$ or $\mu = 0.05$) require a very low optimal complexity ($q \leq 6$).

4.6 Chapter Summary

This chapter provides the complete analytical solution for the design of a planar cantilevered structure with a circular foundation to support a given bending load, using the smallest volume of material required to support the load, while satisfying constraints against material yielding. In the presence of any mass added at the joints (glue, etc.), the minimal mass structure has an optimal complexity. Even without joint mass, the reduction in mass is negligible beyond complexity $q = 5$. We name the discrete truss structure described in Figure 4.3 the *Michell Topology.* The specific properties of the new results are discussed below.

For external forces applied at any or every node

The "complexity" q relates to the number of members ($= q(q+1)$) in the *Michell Topology* ($q = 4$ in Figure 4.3). From node to node, the forces within the structure propagate according to linear recursive equations. Analytic formulas are derived to compute the static equilibrium forces under arbitrary load scenarios.

For external forces applied at only one node ($\mathbf{n}_{00}$)

The properties of such structures are completely characterized as a function of its complexity (q) and the aspect ratio of the structure (ρ^{-1}). The number $\rho^{-1} = r_0/r_q$ represents an aspect ratio for the structure. In Figure 4.3, the radius r_0 is the length of the structure from reference point $\mathbf{0}$ and r_q is the radius of the circular foundation boundary.

Under the assumption that the maximum tensile and compressive stresses are the same, the minimal total material volume subject to a yielding constraint is proportional to the simple quantity $q(\rho^{-1/q} - \rho^{1/q})$. As in [Mic04], in the presence of a single external force applied for bending, the optimal material topology depends on neither material choice nor magnitude of the load. The results give the optimal design for any fixed complexity q, but the minimum volume of material occurs at infinite complexity ($q \rightarrow \infty$).

The Michell Topology is a class 2 tensegrity structure, composed of only axially loaded elements, "sticks and strings". The Michell Topology also has the following features.

The direction of the forces (tension or compression) in every member remains constant, independent of the complexity q or the aspect ratio ρ^{-1}. Furthermore, this unidirectional property of member forces remains independent of the *magnitude* of the external force, depending only on the *direction* of the external force. In fact, for a very large variation in the direction of the external load the members experience unidirectional forces, either tensile or compressive, regardless of the magnitude of the external load. This fact is illustrated in the $q = 4$ example of Figure 4.6, where member forces are *unidirectional* for any load within the shaded area shown. The ability to identify the condition for unidirectionality allows tensegrity design and construction of the optimal bending structure, since conditions for slack strings are clearly identified.

The angle between the connecting tensile and compressive members approaches 90° ($\beta \rightarrow \pi/4$) in the limit as the complexity q approaches infinity. This agrees with the infinitely complex results (filling the space with a material continuum) of Michell [Mic04], where the intersection of tensile and compressive stress lines is always at 90°. In our finite-complexity (discrete) results the angle between tensile and compressive members (2β) is optimized and is always less than 90°. In fact the tangent of the optimal β is simply $\rho^{1/q}$, where $\rho \in [0, 1]$.

Ignoring joint mass, as in the above discussions, the optimal structural complexity is at $q \rightarrow \infty$, yielding the material continuum of Michell. However, even if joint mass is ignored, very low complexity produces a material volume close to the continuum optimum. From Figure 4.12, the optimized material volume is within 7% of the continuum solution when $q \geq 5$ and within 2% when $q \geq 10$.

Practically, increasing complexity q to large values does not yield improved designs, since other neglected issues (such as fabrication errors, material inhomogeneities, ignored fasteners, and glue mass) will likely have more effect than the suboptimality due to the use of a finite q. We quantify this effect by adding to the structural volume objective function a mass (volume) penalty in proportion to the number of joints. Such fabrication effects are characterized in Figure 4.13, which shows that if the joint mass is within 0.1% of the mass of the members being joined, then there is an optimal complexity at $q = 10$. The figure shows that more massive joints lead to less complexity for the optimal structure.

A note on the book cover

The Native American Indian folklore describes a *dreamcatcher* that the Indian would hang outside his tent at night. (See one made by Indians in Figure 4.14.) The good dreams are caught by the *dreamcatcher* and travel to the rim and down the feathers and pass through the similar feathers worn by the sleeper. The bad dreams are trapped at the center of the *dreamcatcher* and are held there until the morning light destroys them. It is interesting to observe that the topology woven by the Indians in the *dreamcatcher* is the same as the Michell Topology in Figures 4.3 and 4.4, repeated to fill the circle.

4.7 Advanced Material

4.7.1 Force Equilibrium at a Generic Node

Multiply (4.13) by the vector $p_{i+k}e^{j(k-i)\phi}$ to obtain

$$p_{i+k}[f_{ik}e^{j\beta} + t_{ki}e^{-j\beta} - f_{i,k-1}e^{j(\beta+\phi)} - t_{k,i-1}e^{-j(\beta+\phi)} + w_{ik}e^{j\theta_{ik}}] = 0. \tag{4.29}$$

This complex equation yields two real equations

$$\begin{bmatrix} \cos\beta & \cos\beta \\ -\sin\beta & \sin\beta \end{bmatrix} \begin{pmatrix} t_{ki} \\ f_{ik} \end{pmatrix} p_{i+k} \\ - \begin{bmatrix} \cos(\beta+\phi) & \cos(\beta+\phi) \\ -\sin(\beta+\phi) & \sin(\beta+\phi) \end{bmatrix} \begin{pmatrix} t_{k,i-1} \\ f_{i,k-1} \end{pmatrix} p_{i+k} + \begin{bmatrix} \cos\theta_{ik} \\ \sin\theta_{ik} \end{bmatrix} w_{ik} = \begin{pmatrix} 0 \\ 0 \end{pmatrix},$$

which are then solved for the two variables $t_{ki}p_{i+k}$ and $f_{ik}p_{i+k}$, leading to (4.14–4.17), where we have used the fact that the lengths p_{i+k} and p_{i+k-1} are related by (4.1).

Figure 4.14: A native American *dreamcatcher*

4.7.2 Proof of Theorem 4.2

A proof of Theorem 4.2 is obtained after an exhaustive inspection of the force equilibrium at all nodes of the Michell Topology. For $i = k = 0$ this was already done in Example 4.1. Further assume that $0 < \beta < \theta_{00} < \pi - \beta$ so that $t_{00} > 0$ and $f_{00} < 0$. The opposite case is handled similarly.

Now move to node $\mathbf{n}_{01}$. In the diagram of Figure 4.5, $\mathbf{w}_{01} = \mathbf{0}$ by assumption, hence $t_{1,-1} = 0$ because $\mathbf{n}_{01}$ is at the lower boundary of the truss, and $f_{00} = |f_{00}| > 0$, where the magnitude of f_{00} is the one computed in node $\mathbf{n}_{00}$. At this point, one can repeat the analysis of Example 4.1 for a fictitious force $\mathbf{w}_{01} = |f_{00}|\mathbf{m}_{00}$. Note that because the truss is such that $a < 1$ one has $\beta < \theta_{01} = \phi + \beta < \pi - \beta$ leading to the conclusion that $t_{01} > 0$ and $f_{10} < 0$. The same idea, applied to all other nodes at the lower boundary, e.g., $\mathbf{n}_{0i}$, $i \leq q$, leads to the conclusion that

$$t_{0i} > 0 \quad \text{and} \quad f_{i0} < 0 \quad \text{for all} \quad i \leq q.$$

The analysis at the node $\mathbf{n}_{10}$ is similar. Following the diagram of Figure 4.5, repeat the analysis of Example 4.1 for a fictitious force $\mathbf{w}_{10} = -|t_{00}|\overline{\mathbf{m}}_{00}$. Once again the fact that the truss is such that $a < 1$ leads to $\pi - \beta > \theta_{10} = \pi - (\phi + \beta) > \beta$ so that $t_{10} > 0$ and $f_{01} < 0$. The same idea, applied to all nodes at the upper boundary, e.g., $\mathbf{n}_{k0}$, $k \leq q$, proves that

$$t_{k0} > 0 \quad \text{and} \quad f_{0k} < 0 \quad \text{for all} \quad k \leq q.$$

The next node to be analyzed is $\mathbf{n}_{11}$. Follow the diagram of Figure 4.5 and repeat the analysis of Example 4.1 for a fictitious force $\mathbf{w}_{11} = -|t_{10}|\overline{\mathbf{m}}_{10} + |f_{10}|\mathbf{m}_{10}$. Using again the fact that $a < 1$ one concludes that $\beta < \theta_{11} < \pi - \beta$, yielding $t_{11} > 0$ and $f_{11} < 0$.

The last type of node that needs to be analyzed is an interior node that is not on the axis $\mathbf{0} - \mathbf{n}_{00}$. For instance, at node $\mathbf{n}_{12}$ create the fictitious force $\mathbf{w}_{12} = -|t_{20}|\overline{\mathbf{m}}_{20} + |f_{11}|\mathbf{m}_{11}$. Again $a < 1$ leads to $\beta < \theta_{12} < \pi - \beta$, yielding $t_{21} > 0$ and $f_{12} < 0$. This process can be repeated sequentially in all remaining nodes of the structure, yielding always an analysis similar to one of the above types of nodes already analyzed. The conclusion is that

$$t_{ki} > 0 \quad \text{and} \quad f_{ik} < 0 \quad \text{for all} \quad i \leq q, \quad k \leq i.$$

A similar procedure where $-\beta > \theta_{00} > \beta - \pi$ leads to

$$t_{ki} < 0 \quad \text{and} \quad f_{ik} > 0 \quad \text{for all} \quad i \leq q, \quad k \leq i,$$

which concludes this proof.

4.7.3 Michell Topologies Under a Single Bending Load

Using the notation introduced in Section 4.2.2 we can express J_q as

$$J_q = (\bar{\lambda} + \bar{\gamma}) \sum_{\alpha=0}^{q-1} \mathbf{y}_\alpha^T \mathbf{x}_\alpha,$$

where $\mathbf{y}_\alpha$ is a vector with the following particular structure:

$$\mathbf{y}_\alpha = \begin{pmatrix} 1 \\ -1 \\ \vdots \\ 1 \\ -1 \end{pmatrix} \in \mathbb{R}^{2(\alpha+1)}.$$

The following result is now needed.

Lemma 4.1 *Let $\mathbf{y}_\alpha$ be defined as above and let $\mathbf{A}_\alpha$ be as defined by (4.20). Then*

$$\mathbf{y}_{\alpha+1}^T \mathbf{A}_\alpha = \mathbf{y}_\alpha^T, \tag{4.30}$$

for any $\alpha = 0, \ldots, q-1$.

Proof: Expand $\mathbf{y}_{\alpha+1}^T \mathbf{A}_\alpha$ and use the fact that $g + h = 2$. □

With the above result in mind we will obtain the interesting recursive formula

$$\begin{aligned} \mathbf{y}_\alpha^T \mathbf{x}_\alpha &= \mathbf{y}_\alpha^T (\mathbf{A}_{\alpha-1}\mathbf{x}_{\alpha-1} + \mathbf{B}_{\alpha-1}\mathbf{u}_{\alpha-1}) \\ &= \mathbf{y}_{\alpha-1}^T \mathbf{x}_{\alpha-1} + \mathbf{y}_\alpha^T \mathbf{B}_{\alpha-1}\mathbf{u}_{\alpha-1}. \end{aligned}$$

Noticing that in the case of a single bending load $\mathbf{w}_{00} = \mathbf{w}$ we have $\mathbf{u}_\alpha = \mathbf{0}$ for all $\alpha = 1, \ldots, q-1$ we conclude that

$$J_q = (\bar{\lambda} + \bar{\gamma}) \sum_{\alpha=0}^{q-1} \mathbf{y}_\alpha^T \mathbf{x}_\alpha = q\, J_1.$$

4.7.4 Michell Topologies with Minimum Material Volume Under a Single Bending Load

Solve for β as follows:

$$\cos\phi + \frac{1}{\tan\beta}\sin\phi = \frac{\sin(\beta+\phi)}{\sin\beta} = \rho^{-1/q} \quad \Rightarrow \quad \tan\beta = \frac{\sin\phi}{\rho^{-1/q} - \cos\phi}. \tag{4.31}$$

Notice that under the above constraint

$$J_q' = \frac{q\sin\phi}{\sin^2\beta}\frac{\sin\beta}{\sin(\beta+\phi)} = \frac{q\,\sin\phi}{\sin^2\beta}\rho^{1/q} = q\,\rho^{1/q}\sin\phi\left(1 + \frac{1}{\tan^2\beta}\right),$$

so that using the expression for $\tan\beta$ computed in (4.31) we obtain

$$J_q' = q\,\rho^{1/q}\sin\phi\left(1 + \frac{\left(\rho^{-1/q} - \cos\phi\right)^2}{\sin^2\phi}\right) = \frac{q}{\sin\phi}\left(\rho^{-1/q} + \rho^{1/q} - 2\cos\phi\right).$$

Because J_q' is a function of ϕ and q, we therefore proceed by determining the optimal value of ϕ that minimizes J_q' as a function of q. This is obtained by solving

$$\frac{d}{d\phi} J_q' = \frac{q}{\sin^2\phi}\left[2 - (\rho^{-1/q} + \rho^{1/q})\cos\phi\right] = 0.$$

One can verify that the optimal choice of ϕ is the one given in (4.26), which is repeated here for convenience,

$$\phi = \phi^* := \arccos\left(\frac{2}{\rho^{-1/q} + \rho^{1/q}}\right)$$

is the unique solution to the above equation in the range $[0, \pi/2]$.

Alternative expressions for ϕ^* can be obtained by noticing that

$$\sin\phi^* = \sqrt{1 - \cos^2\phi^*} = \frac{\rho^{-1/q} - \rho^{1/q}}{\rho^{-1/q} + \rho^{1/q}}, \qquad \tan\phi^* = \frac{\rho^{-1/q} - \rho^{1/q}}{2}, \tag{4.32}$$

which are valid for $\rho \in [0, 1]$ and $q \geq 1$. The above expressions will be useful later.

The exact value of ρ for which $q\,\phi^* = \pi$ as a function of q can be computed by solving the equation

$$q\,\phi^* = q\arctan\left(\frac{\rho^{-1/q} - \rho^{1/q}}{2}\right) = \pi$$

for ρ or, equivalently, by solving the simpler equation

$$\frac{1}{x} - x = 2\tan\left(\frac{\pi}{q}\right) .$$

for $x = \rho^{1/q}$. For $q \geq 2$ the above equation is solved by

$$\rho_\pi = x^q = \left(\sqrt{1 + \tan(\pi/q)^2} - \tan(\pi/q)\right)^q = \left(\frac{1}{\cos(\pi/q)} - \tan(\pi/q)\right)^q,$$

which is (4.27).

Because $\tan\beta$ is a function of ϕ given by (4.31) we can also determine the optimal value of β from

$$\tan\beta^* = \frac{\sin\phi^*}{\rho^{-1/q} - \cos\phi^*} = \frac{\rho^{-1/q} - \rho^{1/q}}{\rho^{-1/q}(\rho^{-1/q} + \rho^{1/q}) - 2} = \rho^{1/q}, \tag{4.33}$$

which yields (4.29). The invariant property (4.28) follows from (4.32) and (4.33).

4.7.5 The Limit as q Goes to ∞

An analytic expression for the boundary of the continuous spiral can be obtained by noticing that nodes on a discrete Michell Spiral satisfy

$$\mathbf{n}_{i+1,k} = a\, e^{j\phi} \mathbf{n}_{ik}.$$

For both families of discrete spirals we can obtain the corresponding continuum spiral by calculating the normalized secant vectors:

$$\mathbf{v}_{i+1,k} = \frac{\mathbf{n}_{i+1,k} - \mathbf{n}_{i,k}}{p_{i+k}} = \frac{(a\, e^{j\phi} - 1)}{p_{i+k}} \mathbf{n}_{i,k} = \frac{(a\, e^{j\phi} - 1)}{c} \frac{\mathbf{n}_{i,k}}{r_{i+k}}.$$

A simple calculation reveals that

$$\frac{(a\, e^{j\phi} - 1)}{c} = \frac{\sin(\beta + \phi)}{\sin \phi} \left(\frac{\sin \beta}{\sin(\beta + \phi)} e^{j\phi} - 1 \right) = -\cos \beta + j \sin \beta = -e^{-j\beta},$$

which is independent of ϕ. From differential geometry [Str61], a continuous curve can be obtained by solving the differential equation

$$\dot{\mathbf{x}}(s) = \mathbf{v}(s) = -e^{-j\beta} \frac{\mathbf{x}(s)}{\|\mathbf{x}(s)\|},$$

where the "dot" denotes differentiation with respect to the arc-length s. By parametrizing the curve in polar coordinates $\mathbf{x}(s) = r(s) e^{j\psi(s)}$ we obtain the complex differential equation

$$\dot{\mathbf{x}}(s) = \left(\dot{r}(s) + j\, r(s)\, \dot{\psi}(s) \right) e^{j\psi(s)} = -e^{-j\beta} e^{j\psi(s)}.$$

This complex differential equation is equivalent to the pair of real differential equations

$$\dot{r}(s) = -\cos \beta, \qquad r(s)\, \dot{\psi}(s) = \sin \beta.$$

One can verify that the solution to these differential equations are

$$r(s) = A - s \cos \beta, \qquad \psi(s) = B - \tan \beta \ln(s \cos \beta - A),$$

where A and B are constants to be determined. For instance, if $\psi(0) = 0$, then

$$A = r(0), \qquad B = \tan \beta \ln(-r(0)),$$

so that

$$r(s) = r(0) - s \cos \beta, \qquad \psi(s) = -\tan \beta \ln \left(\frac{r(0) - s \cos \beta}{r(0)} \right).$$

A more familiar parametrization for the above curve is obtained using $\psi(s)$ as the independent parameter. Indeed

$$\psi(s) = -\tan\beta \ln\left(\frac{r(s)}{r(0)}\right) \qquad \Longrightarrow \qquad r(s) = r(0)e^{-\psi(s)\cot\beta}.$$

In other words

$$\mathbf{x}(\psi) = r(0)\, e^{-\psi \cot\beta} e^{j\psi},$$

which is the familiar *logarithmic spiral*, the optimal solution obtained by Michell [Mic04]. Note that the conjugate curve is

$$\overline{\mathbf{x}}(\eta) = r(0)\, e^{-\eta \cot\beta} e^{-j\eta},$$

which can be obtained as the limit of the conjugate spirals $\mathbf{n}_{i,k+1} = a\, e^{-j\phi}\mathbf{n}_{ik}$ by solving the differential equation

$$\dot{\overline{\mathbf{x}}}(s') = \overline{\mathbf{v}}(s') = -e^{j\beta}\frac{\overline{\mathbf{x}}(s')}{\|\overline{\mathbf{x}}(s')\|},$$

where s' is the arc-length for the conjugate spiral. Note that for any $r(0) = r$, $r \leq r_0$ that the spiral and its conjugate describe the lines of constant compressive and tensile stress, respectively, depending on the orientation of the external load. Indeed at any intersection point of these stress lines we should have $\overline{\mathbf{x}}(s') = \mathbf{x}(s)$ for some s' and s so that the cosine of the angle between the tangent vectors $\overline{\mathbf{v}}(s')$ and $\mathbf{v}(s)$ is given by

$$\mathrm{Re}\{\overline{\mathbf{v}(s')}\,\overline{\mathbf{v}}(s)\} = \mathrm{Re}\left\{ e^{j\beta}\frac{\overline{\mathbf{x}(s')}}{\|\mathbf{x}(s')\|} e^{j\beta}\frac{\overline{\mathbf{x}}(s)}{\|\overline{\mathbf{x}}(s)\|}\right\} = \mathrm{Re}\{e^{j2\beta}\} = \cos(2\beta),$$

where Re denotes the real part of a complex number. Note that when $\beta = \lim_{q\to\infty}\beta^* = \pi/4$ then these tangent vectors are orthogonal thus satisfying Michell's optimality condition [Mic04].

Chapter 5

Analysis of Tensegrity Dynamics

Throughout this chapter we construct dynamic models in the form of ordinary differential equations for tensegrity structures. We make the following assumptions:

a) Rods are rigid, thin, and long and so rotational motion about the longitudinal axis can be neglected.

b) Strings are massless elastic elements with Hookean (linear) behavior only when in tension.

c) The connectivity of the structure is fixed.

These assumptions reflect tensegrity structures where the rods are massive and stiff, here approximated as rigid, as compared with a network of lightweight, elastic strings. Herein we are motivated by the network approach in [Ske05]. We first study the dynamics of a single rod.

5.1 Vectors and Notation

In dynamics a vector was conceived as an entity that has magnitude and direction in three-dimensional space. This concept was introduced by Gibbs (see [Hug86]). In the more modern linear algebra, the axiomatic definitions of a vector allow the treatment of an n-dimensional space, but the concepts of inner products and outer products in linear algebra do not exactly match the concepts of dot and cross products of the dynamics literature. Such distinctions should be made clear. Let $\vec{r}$ be the label we use to represent a (Gibbs) vector in the three-dimensional (non-relativistic) space. This vector is defined independently from any basis system or frame of reference. While

R.E. Skelton, M.C. de Oliveira, *Tensegrity Systems*,
DOI 10.1007/978-0-387-74242-7_5, © Springer Science+Business Media, LLC 2009

the vector is not *defined* by any reference frame, this vector can be *described* in any chosen reference frame. With that in mind we define the following entities.

Definition 5.1 (Dextral Set) *The set of vectors $\vec{e}_i$, i = 1,2,3, form a* dextral set *if the dot products satisfy $\vec{e}_i \cdot \vec{e}_j = \delta_{ij}$ (where δ_{ij} is a Kronecker delta) and the cross products satisfy $\vec{e}_i \times \vec{e}_j = \vec{e}_k$, where the indices i,j,k form the cyclic permutations, i,j,k = 1,2,3 or 2,3,1, or 3,1,2.*

Definition 5.2 (Vectrix) *Let $\vec{e}_i, i = 1, 2, 3$, define a* dextral set *of unit vectors fixed in an inertial frame, and define the* vectrix *$\vec{E}$ by $\vec{E} = \begin{bmatrix} \vec{e}_1 & \vec{e}_2 & \vec{e}_3 \end{bmatrix}$.*

The item $\vec{E}$ is called a *vectrix*, since it is an 1×3 array of the three horizontally stacked items $\vec{e}_i, i = 1, 2, 3$ (the dextral set). Hence, these arrays $\vec{E}$ contain Gibbs vectors $\vec{e}_i$, so they are not matrices. Neither is the 1×3 item $\vec{E}$ a *vector* in the sense of linear algebra. Hence the label *vectrix*, coined by Peter Hughes [Hug86].

Now consider two reference frames, described by the dextral sets (vectrices $\vec{E}$ and $\vec{X}$), where the coordinate transformation between these two frames is described by the 3×3 direction cosine matrix $\mathbf{X}^E$ (orthonormal) so that $\vec{X} = \vec{E}\mathbf{X}^E$, $\mathbf{X}^{E^T}\mathbf{X}^E = \mathbf{I}_3$. Let the three-dimensional column vectors $\mathbf{r}^X$ and $\mathbf{r}^E$ describe the components of the same vector $\vec{r}$ in the two reference frames $\vec{X}$ and $\vec{E}$, respectively. That is,

$$\vec{r} = \vec{E}\mathbf{r}^E = \vec{X}\mathbf{r}^X. \tag{5.1}$$

Hence, if we wish to describe the relationship between the components of the same vector $\vec{r}$, described in two different reference frames, then

$$\vec{X} = \vec{E}\mathbf{X}^E, \qquad \vec{r} = \vec{X}\mathbf{r}^X = \vec{E}\,\mathbf{X}^E\mathbf{r}^X. \tag{5.2}$$

After all terms in an equation are written in the same basis, then the chosen basis (vectrix $\vec{E}$ in this chapter) can be dropped, yielding

$$\mathbf{r}^E = \mathbf{X}^E\mathbf{r}^X. \tag{5.3}$$

The item labeled $\vec{r}$ is a "Gibbs vector", and the items labeled $\mathbf{r}^X$ and $\mathbf{r}^E$ are "vectors" in the spirit of the linear vector spaces of linear algebra, where we use the notation $\mathbf{r}^X, \mathbf{r}^E \in \mathbb{R}^3$ to denote that the items $\mathbf{r}^X$ and $\mathbf{r}^E$ live in a real three-dimensional space. However, the items $\mathbf{r}^X$ and $\mathbf{r}^E$ provide no useful information unless we have previously specified the frames of reference $\vec{X}$ and $\vec{E}$ for these quantities.

The above discussions on notation is for those familiar with traditional literature on rigid body dynamics. However, unlike many problems in aerospace, where multiple coordinate frames are utilized (one fixed in each body), this chapter uses only one coordinate frame (the inertial frame, described by the vectrix $\vec{E}$) to describe all vectors. Hence, one could then shorten the notation for convenience. Instead of the proper notation of a Gibbs vector $\vec{n}_i = \vec{E}\mathbf{n}_i^E$, we will simplify the notation to $\mathbf{n}_i^E = \mathbf{n}_i$ and write $\vec{n}_i = \vec{E}\mathbf{n}_i$, where $\mathbf{n}_i^T = \begin{bmatrix} n_{i_x}, & n_{i_y}, & n_{i_z} \end{bmatrix}$ describes the components of the vector $\vec{n}_i$ in coordinates $\vec{E}$. Hence the only difference between the Gibbs vector $\vec{n}_i$ and the three-dimensional array of its components $\mathbf{n}_i$ is that the frame is specified a priori, $\vec{E}$, and in this chapter, all other vectors that might be mentioned are referenced to the same frame $\vec{E}$.

Hence, for the given basis $\vec{E}$, the dot product and the components of the cross product of any two Gibbs vectors $\vec{b} = \vec{E}\mathbf{b}$ and $\vec{f} = \vec{E}\mathbf{f}$ can be written as

$$\vec{b} \cdot \vec{f} = \mathbf{b}^T\mathbf{f}, \tag{5.4}$$

$$\vec{b} \times \vec{f} = \vec{E}\tilde{\mathbf{b}}\,\mathbf{f}, \qquad \tilde{\mathbf{b}} = \begin{bmatrix} 0 & -b_3 & b_2 \\ b_3 & 0 & -b_1 \\ -b_2 & b_1 & 0 \end{bmatrix}. \tag{5.5}$$

Since we are committed to the same reference frame $\vec{E}$ throughout the chapter, we wish not to burden the notation with the explicit notation of the reference frame. So, by a slight abuse of vector notation, in lieu of the more accurate notation of the cross product, $\vec{b} \times \vec{f} = \vec{E}\tilde{\mathbf{b}}\mathbf{f}$, we will simply write $\mathbf{b} \times \mathbf{f} = \tilde{\mathbf{b}}\mathbf{f}$.

5.2 Dynamics of a Single Rigid Rod

We start by defining some important quantities associated with the dynamics of the single rigid rod in three-dimensional space as illustrated in Figure 5.1. This rod has mass $m > 0$ and length $\ell > 0$ with extreme points $\mathbf{n}_j$, $\mathbf{n}_i \in \mathbb{R}^3$, hence $\|\mathbf{n}_j - \mathbf{n}_i\| = \ell$. We often make use of the normalized rod vector

$$\mathbf{b} = \ell^{-1}(\mathbf{n}_j - \mathbf{n}_i), \qquad \|\mathbf{b}\| = 1. \tag{5.6}$$

Any point in the rod can be located by the vector

$$\mathbf{v}(\mu) = \mu\mathbf{n}_j + (1-\mu)\mathbf{n}_i, \tag{5.7}$$

where $\mu \in [0, 1]$. Let $\rho(\mu) \geq 0$ be a density function defined on the interval $\mu \in [0, 1]$ which describes the mass density along the rod, that is,

$$m = \int_0^1 \rho(\mu)\, d\mu > 0.$$

In this section we describe the position of the rod by means of the configuration vector

$$\mathbf{q} = \begin{pmatrix} \mathbf{r} \\ \mathbf{b} \end{pmatrix} \in \mathbb{R}^6, \tag{5.8}$$

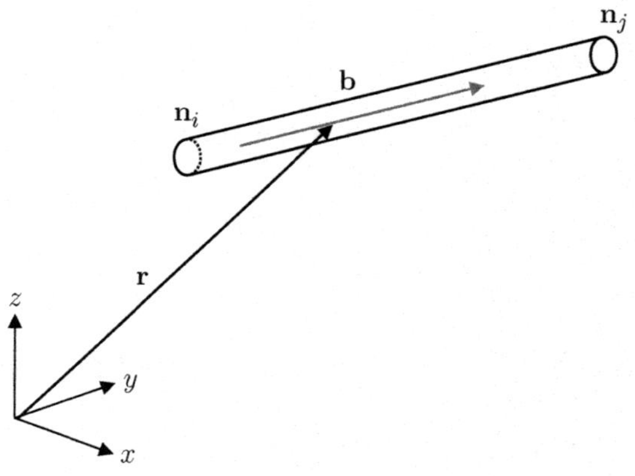

Figure 5.1: Illustration of a rigid rod with its configuration in $\mathbb{R}^3$. The vectors $\mathbf{r}$ and $\mathbf{b}$ describe the translational and rotational position of the rod, respectively

where $\mathbf{r} = \mathbf{v}(\sigma)$, $\sigma \in [0, 1]$, is any fixed point in the rod. Whenever possible $\mathbf{r}$ will be made to coincide with the center of mass of the rod.

Any point in the rod can be equivalently described as a linear function of the configuration vector:

$$\mathbf{v}(\eta) = \mathbf{r} + \eta\,\mathbf{b} = \begin{bmatrix}\mathbf{I}_3 & \eta\mathbf{I}_3\end{bmatrix}\mathbf{q}, \qquad \eta \in [-\sigma\ell, (1-\sigma)\ell]. \tag{5.9}$$

Note that μ and η are related by $\mu = \sigma + \eta/\ell$. Using η we can compute higher order mass moments around $\mathbf{r}$, the next two of which are

$$f(\sigma) = \ell^{-1}\int_{-\sigma\ell}^{(1-\sigma)\ell} \rho(\sigma + \eta/\ell)\,\eta\,d\eta,$$

$$J(\sigma) = \ell^{-1}\int_{-\sigma\ell}^{(1-\sigma)\ell} \rho(\sigma + \eta/\ell)\,\eta^2\,d\eta > 0.$$

Such moments are associated with two important quantities, the *kinetic energy* and the *angular momentum* of the rod. The kinetic energy of the rod is given by the formula

$$T = \frac{1}{2}\int_0^1 \rho(\mu)\,\dot{\mathbf{v}}(\mu)^T\dot{\mathbf{v}}(\mu)\,d\mu = \frac{1}{2}\,\dot{\mathbf{q}}^T(\mathbf{J}(\sigma)\otimes\mathbf{I}_3)\,\dot{\mathbf{q}}, \tag{5.10}$$

$$\mathbf{J}(\sigma) = \begin{bmatrix} m & f(\sigma) \\ f(\sigma) & J(\sigma)\end{bmatrix} \succeq 0. \tag{5.11}$$

We note that a Kronecker product of two $n \times n$ matrices $\mathbf{A}$ and $\mathbf{B}$, denoted by $\mathbf{A} \otimes \mathbf{B}$, is an $n^2 \times n^2$ matrix composed of $n \times n$ blocks of matrices of the type $A_{ij}\mathbf{B}$.

The angular momentum of the rod about $\mathbf{r}$ is

$$\begin{aligned}\mathbf{h} &= \int_m \rho(\mu)\,(\mathbf{v}(\mu) - \mathbf{r}) \times (\dot{\mathbf{v}}(\mu) - \dot{\mathbf{r}})\, d\mu \\ &= J(\sigma)\,\mathbf{b} \times \dot{\mathbf{b}}, \qquad (5.12)\\ &= J(\sigma)\,\tilde{\mathbf{b}}\,\dot{\mathbf{b}}, \qquad (5.13)\end{aligned}$$

where $\tilde{\mathbf{b}}$ denotes a skew-symmetric matrix composed of the three components of the vector $\mathbf{b}$, as defined in (5.4).

The matrix $\mathbf{J}(\sigma)$ is positive semidefinite because $T \geq 0$ for all $\dot{\mathbf{q}}$. Indeed, for most practical mass distribution functions ρ (see next section), matrix $\mathbf{J}$ will be positive definite ($\mathbf{J} \succ 0$), a property that will be used in the next chapters.

Note that if we choose $\sigma = \int_0^1 \rho(\mu)\mu\, d\mu$ so that $\mathbf{r}$ coincides with the center of mass of the rod then $f(\sigma) = 0$. This leads to the well-known decoupling of the kinetic energy in translational and rotational components in a rigid body described by its center of mass. One should choose to describe a rod by its center of mass whenever possible, with the main exception being the case when constraints are present in points of the rod other than the center of mass. We will illustrate this case later in this book. The next example discusses some useful mass distributions and their properties.

Example 5.1

In most parts of this book we consider rods with mass uniformly distributed along the rod, that is,

$$\rho(\mu) = m\,\ell^{-1}.$$

In this case the mass moments f and J are

$$f(\sigma) = \frac{1}{2} m\ell\,(1 - 2\sigma)\,, \qquad J(\sigma) = \frac{1}{3} m\ell^2\,(1 - 3\sigma + 3\sigma^2)\,,$$

which are functions of σ, hence depends on the choice of the fixed point $\mathbf{r}$. Indeed, in this case the center of mass is the center of the bar, i.e., $\sigma = 1/2$ in which case f and J are familiar

$$f(1/2) = 0, \qquad J(1/2) = \frac{1}{12} m\ell^2.$$

Another familiar choice is when $\mathbf{r}$ coincides with one of the extreme points of the rod, say $\mathbf{r} = \mathbf{n}_i$ ($\sigma = 0$) so that

$$f(0) = \frac{1}{2} m\ell, \qquad J(0) = \frac{1}{3} m\ell^2.$$

Interestingly, for any rod with uniform mass distribution matrix $\mathbf{J}$ is positive definite, i.e., $\mathbf{J} \succ 0$, regardless of σ. Indeed, for any $\sigma \in [0,1]$ the function $J(\sigma)$ has two imaginary roots so that

$$J(\sigma) > 0, \qquad m - \frac{f(\sigma)^2}{J(\sigma)} = \frac{m}{4\,(1 - 3\sigma + 3\sigma^2)} > 0 \qquad \text{for all } \sigma \in [0,1].$$

Using the Schur complement [BGFB94] this implies $\mathbf{J} \succ 0$.

The matrix $\mathbf{S} = \mathbf{A} - \mathbf{B}\mathbf{C}^{-1}\mathbf{D}$ is called a *Schur complement* of the matrix $\mathbf{P}$ if either

$$\mathbf{P} = \begin{bmatrix} \mathbf{A} & \mathbf{B} \\ \mathbf{D} & \mathbf{C} \end{bmatrix} \qquad \text{or} \qquad \mathbf{P} = \begin{bmatrix} \mathbf{C} & \mathbf{D} \\ \mathbf{B} & \mathbf{A} \end{bmatrix}. \tag{5.14}$$

Example 5.2

A mass distribution of interest is that comprised of a number of lumped masses along the rod, i.e.,

$$\rho(\mu) = \sum_{k=1}^{K} m_k \delta(\mu - \mu_k),$$

where $\sum_{k=1}^{K} m_k = m$ and $m_k > 0$, $\mu_k \in [0,1]$ for all $k = 1, \ldots, K$. The quantities f and J, expressed as a function of σ, are

$$f(\sigma) = \ell \sum_{k=1}^{K} m_k(\mu_k - \sigma), \qquad J(\sigma) = \ell^2 \sum_{k=1}^{K} m_k(\mu_k - \sigma)^2.$$

In this case

$$J(\sigma) > 0, \quad m - \frac{f(\sigma)^2}{J(\sigma)} = m - \frac{\left(\sum_{k=1}^{K} m_k(\mu_k - \sigma)\right)^2}{\sum_{k=1}^{K} m_k(\mu_k - \sigma)^2} \geq 0 \quad \text{for all } \sigma \in [0,1].$$

Note that for $K > 1$ and $\mu_k \neq \mu_j$ for at least one $j \neq k$ then $m > f(\sigma)^2/J(\sigma)$. For instance, with $K = 2$

$$m - \frac{f(\sigma)^2}{J(\sigma)} = \frac{\ell^2 m_1 m_2 (\mu_1 - \mu_2)^2}{J(\sigma)} > 0 \qquad \text{for all } \mu_1 \neq \mu_2 \text{ and } \sigma \in [0,1],$$

in which case $\mathbf{J} \succ 0$.

In the absence of constraints to the rod kinematics, such as in class 1 tensegrity structures, we find it convenient to work with a *configuration matrix*

$$\mathbf{Q} = \begin{bmatrix} \mathbf{r} & \mathbf{b} \end{bmatrix} \in \mathbb{R}^{3 \times 2} \tag{5.15}$$

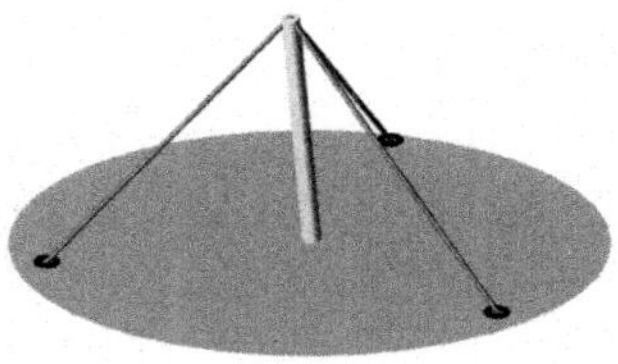

Figure 5.2: A single rod with three strings

as opposed to the configuration vector (5.8). Points on the rod can be described simply as

$$\mathbf{v}(\eta) = \mathbf{Q}\begin{bmatrix}1\\ \eta\end{bmatrix}. \tag{5.16}$$

Compare the above with (5.9).

5.2.1 Nodes as Functions of the Configuration

In dynamics, the node vectors must be expressed as a function of the configuration matrix $\mathbf{Q}$ or the configuration vector $\mathbf{q}$. In the next sections we focus on the configuration matrix $\mathbf{Q}$. The configuration vector $\mathbf{q}$ will be considered when we deal with constrained rods in Section 5.4. One of the major advantages of our approach is that the relationship between the configuration matrix and nodes is linear for all tensegrity structures, as illustrated in the next example.

Example 5.3

Consider the single rod pinned at one end with three strings as illustrated in Figure 5.2. Let $\mathbf{Q}$ be as in (5.15). Because any node $\mathbf{n}_i$ located in a rod is computed through

$$\mathbf{n}_i = \mathbf{Q}\begin{bmatrix}1\\ \eta_i\end{bmatrix}, \qquad i = \{1, 2\},$$

where $\eta_1 = 0$, $\eta_2 = \ell$, we have that

$$\mathbf{N} = \begin{bmatrix}\mathbf{n}_1 & \cdots & \mathbf{n}_5\end{bmatrix} = \mathbf{Q}\begin{bmatrix}1 & 1 & 0 & 0 & 0\\ 0 & \ell & 0 & 0 & 0\end{bmatrix} + \begin{bmatrix}\mathbf{0} & \mathbf{0} & \mathbf{n}_3 & \mathbf{n}_4 & \mathbf{n}_5\end{bmatrix}.$$

In general we should have

$$\mathbf{N} = \mathbf{Q}\,\mathbf{\Psi}^T + \mathbf{Y}, \qquad \mathbf{N}, \mathbf{Y} \in \mathbb{R}^{3\times n}, \qquad \mathbf{\Psi} \in \mathbb{R}^{n\times 2}, \tag{5.17}$$

where $\mathbf{\Psi} \in \mathbb{R}^{n\times 2}$ and $\mathbf{Y} \in \mathbb{R}^{3\times n}$ are constant. The above expression is valid even when more than one rod is considered (see Section 5.3).

5.2.2 String Forces

Forces on the rod are due to the elongation of strings and ground reactions. For simplicity, we assume that the strings are Hookean, as in Section 2.3, and that they are firmly attached to nodes on the rods or on fixed space coordinates. That is, strings are linear force elements with rest length l_i^0 and stiffness k_i. The force vector of the ith string is

$$\mathbf{t}_i := \begin{cases} 0, & \|\mathbf{s_i}\| < l_i^0, \\ -\kappa_i(\|\mathbf{s_i}\| - l_i^0)\|(\mathbf{s_i}/\|\mathbf{s_i}\|), & \|\mathbf{s_i}\| \geq l_i^0, \end{cases} \tag{5.18}$$

where $\mathbf{s}_i$ is a vector in the direction of the ith string. String vectors are linear functions of the nodes of the structure. As in Sections 2.1 and 2.4, assembling a matrix of string vectors and nodes

$$\mathbf{S} = \begin{bmatrix} \mathbf{s}_1 & \cdots & \mathbf{s}_m \end{bmatrix} \in \mathbb{R}^{3\times m}, \quad \mathbf{N} = \begin{bmatrix} \mathbf{n}_1 & \cdots & \mathbf{n}_n \end{bmatrix} \in \mathbb{R}^{3\times n}, \quad \mathbf{S} = \mathbf{N}\mathbf{C}_S^T,$$

where the vector $\mathbf{n}_k$ denotes the kth node in the structure and the string connectivity matrix $\mathbf{C_S} \in \mathbb{R}^{m\times n}$, it follows that

$$\mathbf{T} = -\mathbf{S}\boldsymbol{\Gamma}, \qquad \mathbf{F} = \begin{bmatrix} \mathbf{f}_1 \cdots \mathbf{f}_n \end{bmatrix} = \mathbf{T}\mathbf{C}_S = -\mathbf{N}\mathbf{C}_S^T\boldsymbol{\Gamma}\mathbf{C}_S, \tag{5.19}$$

where we made use of the diagonal matrix $\boldsymbol{\Gamma}$ which contains the force densities

$$\gamma_i := \max\{0, \quad \kappa_i(\|\mathbf{s_i}\| - l_i^0)\|/\|\mathbf{s_i}\|\} \tag{5.20}$$

on its diagonal, as in Sections 2.2 and 2.4. The matrix $\mathbf{F}$ is the matrix of nodal forces.

5.2.3 Generalized Forces and Torques

Equations of motion will be written in terms of the configuration matrix or vector, whereas the forces in the previous section are functions of the nodes. Hence, one needs to express forces in terms of the configuration matrix or vector coordinates. That is, one needs to compute *generalized forces.* As shown at the end of the chapter and because of linearity of (5.17) the matrix of generalized forces is computed as

$$\mathbf{F_Q} = -(\mathbf{Q}\,\boldsymbol{\Psi}^T + \mathbf{Y})\,\mathbf{C}_S^T\,\boldsymbol{\Gamma}\,\mathbf{C}_S\,\boldsymbol{\Psi}. \tag{5.21}$$

A closer look at (5.21) reveals that

$$\mathbf{F_Q} = \begin{bmatrix} \mathbf{f_r} & \mathbf{f_b} \end{bmatrix}, \qquad \mathbf{f_r} = \sum_{i=1}^{n} \mathbf{f}_i, \qquad \mathbf{f_b} = \sum_{i=1}^{n} \eta_i \mathbf{f}_i,$$

where $\mathbf{f_r}$ is simply the sum of all forces applied to the rods and $\mathbf{f_b}$ is related to the sum of the torques on the rod. Indeed

$$\tilde{\mathbf{b}}\,\mathbf{f_b} = \sum_{i=1}^{n} \tau_i, \qquad \tau_i = \eta_i \tilde{\mathbf{b}}\,\mathbf{f}_i.$$

This fact will be used next.

5.2.4 Equations of Motion

Newtonian approach

In this section let us assume that $\mathbf{r}$ coincides with the location of the center of mass of the rod. Then the translation of the center of mass is governed by the equations of motion

$$m\ddot{\mathbf{r}} = \sum_{i=1}^{n} \mathbf{f}_i, \tag{5.22}$$

where $\mathbf{f}_i$ are external forces applied to the rod. Let $\tau = \sum_{i=1}^{n} \tau_i$ be the torques applied to the rod, assuming that the fixed point $\mathbf{r}$ is both the center of mass and center of rotation. Then from Newton's laws $\dot{\mathbf{h}} = \tau$, where $\mathbf{h} = J\,\tilde{\mathbf{b}}\,\dot{\mathbf{b}}$, so that

$$\dot{\mathbf{h}} = J\left(\dot{\tilde{\mathbf{b}}}\,\dot{\mathbf{b}} + \tilde{\mathbf{b}}\,\ddot{\mathbf{b}}\right) = J\,\tilde{\mathbf{b}}\,\ddot{\mathbf{b}}, \tag{5.23}$$

where, for ease of notation, we have omitted the dependence of J on σ. We shall do the same with respect to f from now on. Hence

$$J\,\tilde{\mathbf{b}}\,\ddot{\mathbf{b}} = \sum_{i=1}^{n} \tau_i. \tag{5.24}$$

We must add to these equations a constraint on the length of the rod, $(\ell\mathbf{b})^T(\ell\mathbf{b}) = \ell^2$, or simply $\mathbf{b}^T\mathbf{b} = 1$, as described in (5.6). Differentiating this constraint twice with respect to time yields

$$\mathbf{b}^T\dot{\mathbf{b}} = 0, \qquad \mathbf{b}^T\ddot{\mathbf{b}} + \|\dot{\mathbf{b}}\|^2 = 0. \tag{5.25}$$

Equations (5.23) and (5.25) must be solved simultaneously for $\ddot{\mathbf{b}}$. Note that they are linear in the vector $\ddot{\mathbf{b}}$, yielding the solution for $\ddot{\mathbf{b}}$ (proof at the end of chapter),

$$\ddot{\mathbf{b}} = -(\|\dot{\mathbf{b}}\|/\|\mathbf{b}\|)^2\,\mathbf{b} - \sum_{i=1}^{n} J^{-1}(\tilde{\mathbf{b}}\,\tau_i)/\|\mathbf{b}\|^2. \tag{5.26}$$

We also show at the end of the chapter that with the help of the *projection matrix*

$$\mathbf{P}(\mathbf{b}) := \mathbf{I} - (\mathbf{b}\mathbf{b}^T)/\|\mathbf{b}\|^2, \tag{5.27}$$

we can write

$$\sum_{i=1}^{n} \tilde{\mathbf{b}}\,\tau_i = -\|\mathbf{b}\|^2\mathbf{P}(\mathbf{b})\,\mathbf{f}_{\mathbf{b}}$$

to express the torques in the right-hand side of (5.26) in terms of the generalized force $\mathbf{f}_{\mathbf{b}} = \sum_1^n \eta_i\mathbf{f}_i$ so that,

$$\ddot{\mathbf{b}} = J^{-1}\mathbf{P}(\mathbf{b})\,\mathbf{f}_{\mathbf{b}} - (\|\dot{\mathbf{b}}\|/\|\mathbf{b}\|)^2\,\mathbf{b}. \tag{5.28}$$

Lagrangian approach

In Section 5.4 we will deal with tensegrity structures in which some or all the rods may have kinematic constraints. In such structures it may be advantageous to make $\mathbf{r}$ not coincide with the location of the center of mass of the rod. In such cases, deriving the equations of motion using the momentum approach of the previous section may be unnecessarily complicated. A simpler approach is the use of energy methods, whose full potential we explore in Section 5.4. In the present section we simply rederive the equations of the previous section in order to introduce the reader to energy methods.

Consider the Lagrangian function

$$L = T - V - \frac{J\xi}{2}\left(\mathbf{b}^T\mathbf{b} - 1\right), \tag{5.29}$$

where ξ is the Lagrange multiplier responsible for enforcing the constraint that $\mathbf{b}$ must remain unitary (5.6) and V is some appropriately defined potential function. Assume once again that $\mathbf{r}$ coincides with the location of the center of mass of the rod, i.e., $f = 0$. Following standard derivations as shown at the end of the chapter we arrive at the equations of motion

$$m\,\ddot{\mathbf{r}} = \mathbf{f_r}, \qquad J\,\ddot{\mathbf{b}} = \mathbf{f_b} - J\xi\mathbf{b}, \qquad \mathbf{b}^T\mathbf{b} - 1 = 0, \tag{5.30}$$

where $\mathbf{f_r}$ and $\mathbf{f_b}$ are the vector of generalized forces acting on the rod written in the coordinates $\mathbf{q}$ (see Section 5.2.1).

The difficulty in (5.30) is not solving for $\ddot{\mathbf{b}}$ (which can be done easily because $J > 0$) but avoiding the explicit calculation of the Lagrange multiplier ξ. This can be overcome once again by using the constraint (5.6), as shown in the notes at the end of the chapter, where it is found that

$$\xi = (\|\dot{\mathbf{b}}\|/\|\mathbf{b}\|)^2 + J^{-1}\mathbf{b}^T\mathbf{f_b}/\|\mathbf{b}\|^2. \tag{5.31}$$

Substituting ξ on (5.30) produces the rotational equations of motion

$$\ddot{\mathbf{b}} = J^{-1}\mathbf{P}(\mathbf{b})\,\mathbf{f_b} - (\|\dot{\mathbf{b}}\|/\|\mathbf{b}\|)^2\mathbf{b}, \tag{5.32}$$

where $\mathbf{P}(\mathbf{b})$ is the projection matrix (5.27). Not surprisingly, the above equation is the same as the one previously obtained in (5.28).

5.3 Class 1 Tensegrity Structures

The equations of motion developed in the previous section can be extended to cope with general class 1 tensegrity structures in a fairly straightforward way. Instead of presenting a lengthy and detailed derivation of the equations of motion for general class 1 tensegrity systems, we shall limit ourselves to indicate what are the steps needed to be taken in order to undertake such generalizations based on the material presented so far.

Because in class 1 tensegrity structures no rods touch each other, there exists no extra constraint that should be taken into consideration beyond the ones already considered in Section 5.2. In fact, using the energy approach all that needed to derive equations of motion for a class 1 tensegrity system with K rods is to define the combined Lagrangian

$$L = \sum_{j=1}^{k} L_j,$$

where each L_j is a Lagrangian function written for each rod $j = 1, \ldots, k$ as in (5.29) and following the procedure outlined in Section 5.2.4 for enforcing the individual rod constraints and deriving the equations of motion. With that in mind define the configuration matrix

$$\mathbf{Q} = \begin{bmatrix} \mathbf{R} & \mathbf{B} \end{bmatrix} \in \mathbb{R}^{3\times 2k}, \tag{5.33}$$

where

$$\mathbf{R} = \begin{bmatrix} \mathbf{r}_1 & \cdots & \mathbf{r}_k \end{bmatrix}, \qquad \mathbf{B} = \begin{bmatrix} \mathbf{b}_1 & \cdots & \mathbf{b}_k \end{bmatrix} \in \mathbb{R}^{3\times k}. \tag{5.34}$$

Note that in the absence of constraints (5.17) is still valid provided an appropriate matrix $\mathbf{\Psi} \in \mathbb{R}^{n\times 2k}$ is constructed. Likewise, generalized forces are easily computed using (5.21)

$$\mathbf{F_Q} = \begin{bmatrix} \mathbf{F_R} & \mathbf{F_B} \end{bmatrix} \in \mathbb{R}^{3\times 2k}, \tag{5.35}$$

where

$$\mathbf{F_R} = \begin{bmatrix} \mathbf{f}_{\mathbf{r}_1} & \cdots & \mathbf{f}_{\mathbf{r}_k} \end{bmatrix}, \qquad \mathbf{F_B} = \begin{bmatrix} \mathbf{f}_{\mathbf{b}_1} & \cdots & \mathbf{f}_{\mathbf{b}_k} \end{bmatrix} \in \mathbb{R}^{3\times k}. \tag{5.36}$$

The relationship between each column of $\mathbf{F_B}$ and torques is the same as provided in Section 5.2.1.

A surprisingly compact matrix expression for the resulting equations of motion is possible by combining (5.30) and (5.31) as follows:

$$(\ddot{\mathbf{Q}} + \mathbf{Q}\,\mathbf{\Xi})\,\mathbf{M} = \mathbf{F_Q}, \tag{5.37}$$

where

$$\mathbf{M} = \mathrm{diag}[m_1, \ldots, m_k, J_1, \ldots, J_k] \tag{5.38}$$

is a constant matrix and

$$\mathbf{\Xi} = \mathrm{diag}[0, \ldots, 0, \xi_1, \ldots, \xi_k], \tag{5.39}$$

where ξ_j are Lagrange multipliers computed as in (5.31) for each individual bar $\mathbf{b}_j$. The above discussion is summarized in the next theorem.

Theorem 5.1 *Consider an unconstrained class 1 tensegrity system with k rigid fixed length rods. Define the* configuration matrix (5.33)

$$\mathbf{Q} = \begin{bmatrix} \mathbf{R} & \mathbf{B} \end{bmatrix} \in \mathbb{R}^{3\times 2k},$$

where the columns of $\mathbf{R}$ *describe the center of mass of the k rods and the columns of* $\mathbf{B}$ *describe the rod vectors. The* $\mathbf{\Psi} \in \mathbb{R}^{n\times 2k}$ *and* $\mathbf{Y} \in \mathbb{R}^{3\times n}$ *are constant matrices that relate the $n \geq 2k$ nodes of the structure with the configuration matrix through* (5.17)

$$\mathbf{N} = \mathbf{Q}\,\mathbf{\Psi}^T + \mathbf{Y}, \qquad \mathbf{N}, \mathbf{Y} \in \mathbb{R}^{3\times n}, \qquad \mathbf{\Psi} \in \mathbb{R}^{n\times 2}.$$

The dynamics of such unconstrained class 1 *tensegrity systems satisfy* (5.37)

$$(\ddot{\mathbf{Q}} + \mathbf{Q}\,\mathbf{\Xi})\,\mathbf{M} = \mathbf{F_Q},$$

where

$$\mathbf{M} = \mathrm{diag}[m_1, \ldots, m_k, J_1, \ldots, J_k], \qquad \mathbf{\Xi} = \mathrm{diag}[0, \ldots, 0, \xi_1, \ldots, \xi_k].$$

The Lagrange multipliers ξ_i, $i = 1, \ldots, k$, *are computed by*

$$\xi_i = (\|\dot{\mathbf{b}}_i\|/\|\mathbf{b}_i\|)^2 + J_i^{-1}\mathbf{b}_i^T\mathbf{f}_{\mathbf{b}_i}/\|\mathbf{b}_i\|^2, \tag{5.40}$$

where $\mathbf{f}_{\mathbf{b}_i}$ *are columns of the matrix* $\mathbf{F_B}$ *which is part of the* matrix of generalized forces

$$\mathbf{F_Q} = \begin{bmatrix} \mathbf{F_R} & \mathbf{F_B} \end{bmatrix} \in \mathbb{R}^{3\times 2k},$$

which is computed by (5.21)

$$\mathbf{F_Q} = [\mathbf{W} - (\mathbf{Q}\,\mathbf{\Psi}^T + \mathbf{Y})\,\mathbf{C}_S^T\,\mathbf{\Gamma}\,\mathbf{C}_S\,]\mathbf{\Psi},$$

where $\mathbf{C}_S$ *is the* string connectivity matrix, *and the external force acting on node* $\mathbf{n}_i$ *is* $\mathbf{w}_i$, *and the matrix of all such external forces is*

$$\mathbf{W} = \begin{bmatrix} \mathbf{w}_1 & \mathbf{w}_2 & \cdots & \mathbf{w}_{2k} \end{bmatrix}. \tag{5.41}$$

By parametrizing the configuration in terms of the components of vectors, the usual transcendental nonlinearities involved with the use of angles, angular velocities, and coordinate transformations are avoided. Indeed, the absence of trigonometric functions in this formulation leads to a simplicity in the analytical form of the dynamics. This might facilitate more efficient numerical solutions of the differential equations (simulations) and the design of control laws. Actually, the simplicity of the structure of these equations (5.37) is partly due to the use of the matrix form and partly due to the enlarged space in which the dynamics are described. The actual degrees of freedom for each rod is 5, whereas the model (5.37) has as many equations as required for 6 degrees of freedom for each rod. That is, the equations are a non-minimal realization of the dynamics. The *mathematical structure* of the equations are simple, however. This will allow much easier integration of structure and control design, since the control variables (string force densities) appear linearly, and the simple structure of the nonlinearities can be exploited in later control investigations.

Example 5.4

Consider the tensegrity prism depicted in Figure 5.3. This structure has 6 nodes, 3 rods, and 12 strings. Let the node matrix be

$$\mathbf{N} = \begin{bmatrix} \mathbf{n}_1 & \mathbf{n}_2 & \mathbf{n}_3 & \mathbf{n}_4 & \mathbf{n}_5 & \mathbf{n}_6 \end{bmatrix},$$

where each pair of nodes is a pair of bottom and top nodes on a rod. That is,

$$\mathbf{B} = \begin{bmatrix} \ell_1^{-1}(\mathbf{n}_1 - \mathbf{n}_2) & \ell_2^{-1}(\mathbf{n}_3 - \mathbf{n}_4) & \ell_3^{-1}(\mathbf{n}_5 - \mathbf{n}_6) \end{bmatrix}.$$

Assuming that the mass m_j of the jth rods is uniformly distributed then the center of each rods is its center of mass

$$\mathbf{R} = \frac{1}{2} \begin{bmatrix} \mathbf{n}_1 + \mathbf{n}_2 & \mathbf{n}_3 + \mathbf{n}_4 & \mathbf{n}_5 + \mathbf{n}_6 \end{bmatrix}.$$

The nodes can be retrieved from the configuration matrix $\mathbf{Q} = \begin{bmatrix} \mathbf{R} & \mathbf{B} \end{bmatrix}$ through (5.17) with

$$\mathbf{\Psi} = \begin{bmatrix} 1 & 1 & 1 & 1 & 1 & 1 \\ \ell_1/2 & -\ell_1/2 & \ell_2/2 & -\ell_2/2 & \ell_3/2 & -\ell_3/2 \end{bmatrix},$$

and, because of the uniform mass distribution and the choice of $\mathbf{R}$, we have that $f_j = 0$ and

$$J_j = \frac{1}{12} m_j \ell_j^2, \qquad j = \{1, 2, 3\}.$$

The string connectivity is

$$\mathbf{C}_S = \begin{bmatrix} 1 & 0 & -1 & 0 & 0 & 0 \\ 0 & 0 & 1 & 0 & -1 & 0 \\ -1 & 0 & 0 & 0 & 1 & 0 \\ 0 & 1 & 0 & -1 & 0 & 0 \\ 0 & 0 & 0 & 1 & 0 & -1 \\ 0 & -1 & 0 & 0 & 0 & 1 \\ 0 & 1 & -1 & 0 & 0 & 0 \\ 0 & 0 & 0 & 1 & -1 & 0 \\ -1 & 0 & 0 & 0 & 0 & 1 \end{bmatrix}.$$

With this information one can write the equations of motion (5.37).

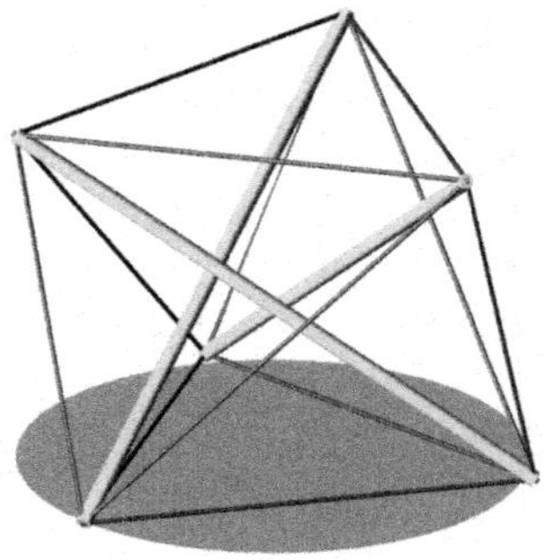

Figure 5.3: A class 1 tensegrity prism with 3 rods and 12 strings

5.4 Constrained Class 1 Tensegrity Structures

We now consider class 1 tensegrity structures in which nodes in some of the rods may have linear kinematic constraints due to its interaction with the environment. Still no rods touch each other. In such cases, it may be advantageous to work with a reduced configuration vector as opposed to our oversized configuration matrix $\mathbf{Q}$, since the latter might not be well defined, as in the next example.

Example 5.5

Let the z-coordinate of node $\mathbf{r} = \mathbf{v}(0)$ of Example 5.3 be constrained to stay at xy-plane, i.e., $\mathbf{r}_z = 0$. Define the *reduced* configuration vector

$$\mathbf{q} = \begin{pmatrix} \mathbf{r}_x \\ \mathbf{r}_y \\ \mathbf{b} \end{pmatrix}.$$

In this case, the relationship between the configuration vector $\mathbf{q}$ and the nodes is of the form

$$\mathbf{n} = \mathbf{\Phi}\,\mathbf{q} + \mathbf{y}. \tag{5.42}$$

Note that when (5.17) holds then (5.42) is obtained from (5.17) by vectorization with $\mathbf{n} = \operatorname{vec}\mathbf{N}$, $\mathbf{y} = \operatorname{vec}\mathbf{Y}$, and $\mathbf{\Phi} = \mathbf{\Psi} \otimes \mathbf{I}_3$. In general, $\mathbf{\Phi} \neq \mathbf{\Psi} \otimes \mathbf{I}_3$, as in the next example. As shown at the end of the chapter, with external forces $\mathbf{w}_i$ added to each node (where the total external vector of forces is $\mathbf{w}$), generalized forces are computed as

$$\mathbf{f}_\mathbf{q} = \mathbf{\Phi}^T\mathbf{w} - \mathbf{\Phi}^T\left(\mathbf{C}_S^T\,\mathbf{\Gamma}\,\mathbf{C}_S \otimes \mathbf{I}_3\right)\left(\mathbf{\Phi}\,\mathbf{q} + \mathbf{y}\right). \tag{5.43}$$

Example 5.6

Consider the reduced configuration vector of Example 5.5. For the same configuration of strings as in Example 5.3 we have

$$\mathbf{\Phi} = \begin{bmatrix} \mathbf{E} & \mathbf{0} \\ \mathbf{E} & \ell \mathbf{I}_3 \\ \mathbf{0} & \mathbf{0} \\ \mathbf{0} & \mathbf{0} \\ \mathbf{0} & \mathbf{0} \end{bmatrix}, \qquad \mathbf{E} = \begin{bmatrix} \mathbf{I}_2 \\ \mathbf{0} \end{bmatrix}, \qquad \mathbf{y} = \begin{bmatrix} \mathbf{0} \\ \mathbf{0} \\ \mathbf{n}_3 \\ \mathbf{n}_4 \\ \mathbf{n}_5 \end{bmatrix}. \tag{5.44}$$

Note that $\mathbf{\Phi} \neq \mathbf{\Psi} \otimes \mathbf{I}_3$.

5.4.1 Single Constrained Rigid Rod

Let $\mathbf{r} = \mathbf{v}(\sigma)$ for some $\sigma \in [0, 1]$ be a fixed point in the rod which may not coincide with the rod's center of mass and is subject to the linear constraint

$$\mathbf{D}\,\mathbf{r} = \bar{\mathbf{r}}, \tag{5.45}$$

where $r = \text{rank}(\mathbf{D}) < 3$ and $\bar{\mathbf{r}} \in \mathbb{R}^r$ constant. Let $\mathbf{E} \in \mathbb{R}^{3\times(3-r)}$ be an orthonormal matrix, i.e., $\mathbf{E}^T\mathbf{E} = \mathbf{I}$, such that $\mathbf{DE} = 0$. Then all solutions to (5.45) are parametrized by

$$\mathbf{r} = \mathbf{D}^{\dagger}\bar{\mathbf{r}} + \mathbf{E}\mathbf{z}, \tag{5.46}$$

where $\mathbf{z} \in \mathbb{R}^{3-r}$. Define the *reduced configuration vector*

$$\mathbf{q} = \begin{pmatrix} \mathbf{z} \\ \mathbf{b} \end{pmatrix}. \tag{5.47}$$

Example 5.7

In Example 5.5 we have $\mathbf{r} = \mathbf{v}(0)$ with

$$\mathbf{D} = \begin{bmatrix} 0 & 0 & 1 \end{bmatrix}, \qquad \bar{\mathbf{r}} = 0.$$

Then

$$\mathbf{D}^{\dagger} = \begin{bmatrix} 0 \\ 0 \\ 1 \end{bmatrix}, \qquad \mathbf{E} = \begin{bmatrix} \mathbf{I}_2 \\ \mathbf{0} \end{bmatrix}.$$

The particular case when $r = \text{rank}(\mathbf{D}) = 3$ is handled by defining the reduced configuration vector as

$$\mathbf{q} = \mathbf{b}, \tag{5.48}$$

because $\mathbf{r} = \mathbf{D}^{-1}\bar{\mathbf{r}}$ is simply a constant.

The above discussion also provides clues on how to compute matrices $\boldsymbol{\Phi}$ and $\mathbf{y}$ in Section 5.2.1. For instance, after one has computed $\boldsymbol{\Psi}$ and $\mathbf{Y}$ such that

$$\mathbf{N} = \mathbf{Q}\,\boldsymbol{\Psi}^T + \mathbf{Y},$$

where $\mathbf{Q}$ is the "non-reduced" configuration matrix (5.15), it becomes clear that $\mathbf{n}$ and $\mathbf{q}$ should be related through (5.42) where

$$\boldsymbol{\Phi} = (\boldsymbol{\Psi} \otimes \mathbf{I}_3)\begin{bmatrix} \mathbf{E} & \mathbf{0} \\ \mathbf{0} & \mathbf{I}_3 \end{bmatrix}, \qquad \mathbf{y} = \operatorname{vec}\mathbf{Y} + (\boldsymbol{\Psi} \otimes \mathbf{I}_3)\begin{pmatrix} \mathbf{D}^\dagger \bar{\mathbf{r}} \\ \mathbf{0} \end{pmatrix},$$

because

$$\operatorname{vec}\mathbf{Q} = \begin{bmatrix} \mathbf{r} \\ \mathbf{b} \end{bmatrix} = \begin{bmatrix} \mathbf{D}^\dagger \bar{\mathbf{r}} + \mathbf{E}\,\mathbf{z} \\ \mathbf{b} \end{bmatrix} = \begin{bmatrix} \mathbf{E} & \mathbf{0} \\ \mathbf{0} & \mathbf{I}_3 \end{bmatrix}\mathbf{q} + \begin{pmatrix} \mathbf{D}^\dagger \bar{\mathbf{r}} \\ \mathbf{0} \end{pmatrix}.$$

Equations of motion

Since $\mathbf{r}$ may not be the center of mass, the equations of motion are expected to be more complex than the ones seen so far. This is justified by the extra work that is required to handle the constraint (5.6). As shown at the end of the chapter, the equations of motion for a single constrained rod are of the form

$$\mathbf{M}(\mathbf{q})\,\ddot{\mathbf{q}} + \mathbf{g}(\mathbf{q}, \dot{\mathbf{q}}) = \mathbf{H}(\mathbf{q})\,\mathbf{f}_{\mathbf{q}}, \tag{5.49}$$

where

$$\mathbf{M}(\mathbf{q}) := \begin{bmatrix} m\,\mathbf{I} - f^2 J^{-1}\mathbf{E}^T\mathbf{P}(\mathbf{b})\mathbf{E} & \mathbf{0} \\ f J^{-1}\mathbf{P}(\mathbf{b})\mathbf{E} & \mathbf{I} \end{bmatrix},$$

$$\mathbf{H}(\mathbf{q}) := \begin{bmatrix} \mathbf{I} & -f J^{-1}\mathbf{E}^T\mathbf{P}(\mathbf{b}) \\ \mathbf{0} & J^{-1}\mathbf{P}(\mathbf{b}) \end{bmatrix},$$

$$\mathbf{g}(\mathbf{q}, \dot{\mathbf{q}}) := (\|\dot{\mathbf{b}}\|/\|\mathbf{b}\|)^2 \begin{pmatrix} -f\,\mathbf{E}^T\mathbf{b} \\ \mathbf{b} \end{pmatrix}. \tag{5.50}$$

When dealing with equations of motion of the form (5.49) an issue that arises is that of solving for $\ddot{\mathbf{q}}$ as a function of $\mathbf{q}$ and $\dot{\mathbf{q}}$. This is indeed the case in most cases of interest where $\mathbf{J}$ is positive definite due to the following lemma which is proved at the end of the chapter.

Lemma 5.1 *Let* $\mathbf{J}$ *and* $\mathbf{M}(\mathbf{q})$ *be defined by* (5.11) *and* (5.50), *respectively. If* $\mathbf{J} \succ 0$ *then* $\mathbf{M}(\mathbf{q})$ *is nonsingular for all* $\mathbf{q}$ *such that* $\|\mathbf{b}\| = 1$.

5.4.2 General Class 1 Tensegrity Structures

The equations of motion developed in the previous section can be generalized to cope with general constrained class 1 tensegrity structures as done in Section 5.3.

After defining local configuration vectors $\mathbf{q}_j$, $j = 1, \ldots, K$, we can follow the derivations in the previous section and arrive at the system of differential equations of the form (5.49), that is,

$$\mathbf{M}_j(\mathbf{q}_j)\,\ddot{\mathbf{q}}_j + \mathbf{g}_j(\mathbf{q}_j, \dot{\mathbf{q}}_j) = \mathbf{H}_j(\mathbf{q}_j)\,\mathbf{f}_{\mathbf{q}_j}, \qquad j = 1, \ldots, K, \tag{5.51}$$

where $\mathbf{M}_j$, $\mathbf{g}_j$, and $\mathbf{H}_j$ are as defined in (5.50) for the jth rod.

Example 5.8

Consider again the tensegrity prism of Example 5.4 depicted in Figure 5.3. As before, the mass of the three rods is assumed to be uniformly distributed but this time $\mathbf{r}_j = \mathbf{v}(0)$, for all $j = \{1, 2, 3\}$,

$$\sigma_j = 0, \qquad \Longrightarrow \qquad f_j = \frac{1}{2} m_j \ell_j, \qquad J_j = \frac{1}{3} m_j \ell_j, \qquad j = \{1, 2, 3\},$$

that is, the vectors $\mathbf{r}_j$ all point to one extreme node of each rod. Now set $\mathbf{r}_1$ to be the origin ($\mathbf{r}_1 = \mathbf{0}$) and consider that nodes $\mathbf{r}_2$ and $\mathbf{r}_3$ be constrained as in (5.45) and (5.46) with

$$\bar{\mathbf{r}}_2 = 0, \qquad \mathbf{D}_2 = \begin{bmatrix} 0 & 0 & 1 \end{bmatrix}, \qquad \mathbf{E}_2 = \begin{bmatrix} 1 & 0 \\ 0 & 1 \\ 0 & 0 \end{bmatrix},$$

$$\bar{\mathbf{r}}_3 = \begin{pmatrix} 0 \\ 0 \end{pmatrix}, \qquad \mathbf{D}_3 = \begin{bmatrix} 1 & 0 & 0 \\ 0 & 0 & 1 \end{bmatrix}, \qquad \mathbf{E}_3 = \begin{bmatrix} 0 \\ 1 \\ 0 \end{bmatrix}.$$

The above matrices reflect the fact that the z-coordinate of the bottom node of the second rod ($\mathbf{r}_2$) is set to zero, i.e., it is free to move only in the xy-plane and the x- and z-coordinates of the bottom node of the third rod ($\mathbf{r}_3$) are set to zero, i.e., it can move only on the y-axis. This set of six constraints eliminates the six rigid body modes of the structure. The configuration vector of the system $\mathbf{q} \in \mathbb{R}^{12}$ is then $\mathbf{q} = (\mathbf{b}_1, \mathbf{z}_2, \mathbf{b}_2, \mathbf{z}_3, \mathbf{b}_3)$.

With the data on the above example one can construct the equations of motion (5.51) with the exception of the generalized force vector $\mathbf{f}_{\mathbf{q}_j}$, which should be computed using (5.43) and the data in the following example.

Example 5.9

For the same tensegrity prism depicted in Figure 5.3 of Examples 5.4 and 5.8 let the node matrix $\mathbf{N} \in \mathbb{R}^{3\times 5}$ be

$$\mathbf{N} = \begin{bmatrix} \mathbf{n}_1 & \mathbf{n}_2 & \mathbf{n}_3 & \mathbf{n}_4 & \mathbf{n}_5 \end{bmatrix},$$

with connectivity matrix

$$\mathbf{C}_S = \begin{bmatrix} 0 & 0 & 0 & 0 & -1 & -1 & -1 & -1 & 0 & 0 & 0 & 0 \\ 1 & 0 & 0 & 0 & 1 & 0 & 0 & 0 & -1 & -1 & 0 & 0 \\ 0 & 1 & 0 & 0 & 0 & 1 & 0 & 0 & 0 & 0 & -1 & -1 \\ 0 & 0 & 1 & 0 & 0 & 0 & 1 & 0 & 1 & 0 & 1 & 0 \\ 0 & 0 & 0 & 1 & 0 & 0 & 0 & 1 & 0 & 1 & 0 & 1 \end{bmatrix}^T \in \mathbb{R}^{12\times 5},$$

and the matrix

$$\Phi = \begin{bmatrix} \ell_1\mathbf{I} & \mathbf{0} & \mathbf{0} & \mathbf{0} & \mathbf{0} \\ \mathbf{0} & \mathbf{E}_2 & \mathbf{0} & \mathbf{0} & \mathbf{0} \\ \mathbf{0} & \mathbf{E}_2 & \ell_2\mathbf{I} & \mathbf{0} & \mathbf{0} \\ \mathbf{0} & \mathbf{0} & \mathbf{0} & \mathbf{E}_3 & \mathbf{0} \\ \mathbf{0} & \mathbf{0} & \mathbf{0} & \mathbf{E}_3 & \ell_3\mathbf{I} \end{bmatrix} \in \mathbb{R}^{15\times 12}.$$

Vector $\mathbf{y} \in \mathbb{R}^{15}$ is equal to zero.

5.5 Chapter Summary

The equations of motion for any tensegrity system composed of rods and strings are provided in simple form, to make computation and control design easier. One might argue that the absence of simple equations for the dynamics of tensegrity systems has been a limiting factor to the acceptance of tensegrity in engineering practice.

Axially loaded elements (rods and strings) are used throughout. Two kinds of constraints are treated in the dynamics. The length of rods are constant, and position of any node may be fixed. The main contributions of the chapter include both energy and Newtonian approaches, constrained and unconstrained systems, non-minimal realizations of the constrained dynamics, and finally a new matrix form of the equations in Theorem 5.1.

To obtain equations that are efficient for dynamic simulation, with constraints, there are many debates about which method is more efficient. Here the energy and the Newtonian derivations produce the same equations. This is done without using the classical angular velocity vector, since in our case where a 5 DOF system is modeled by 6 DOF (non-minimal equations), the angular velocity about the long axis of a rod is undefined. By using the vector along the rod as a generalized coordinate, the final equations are devoid of the transcendental functions that complicate the form of the dynamics. Putting these equations in matrix form allow the mathematical structure of the equations to be extremely simple. The motivation for seeking simple structure of the equations is the hope that control laws can be found to exploit the known simple structure of the dynamic model. This hope is high enough, we believe, to justify the non-minimality of the equations. Quite often in mathematics the minimal number of equations are often the most complex in form.

The constraints are treated with and without the use of Lagrange multipliers. Without Lagrange multipliers it is shown that the correct equations are obtained by linear algebra, to obtain a least squares solution which enforces the constraints.

The distribution of members and forces are characterized as networks, where efficient matrix methods simplify the description of forces and connections. A connectivity matrix is introduced that characterizes the topology of all rod to string connections. These network equations, together with a simple characterization of the dynamics of a rigid body, allow efficient forms for the final equations.

By using force densities as the input variable (later to be the control variable) the final equations of motion for the general nonlinear tensegrity system has a bilinear structure (equations are nonlinear in the generalized coordinates, but linear in the string force densities). This will offer great advantage in control design.

5.6 Advanced Material

5.6.1 Dynamics of a Single Rigid Rod

Most of the quantities defined in Section 5.2 can be visualized directly from Figure 5.1, from where (5.9) follows. The kinetic energy formula comes from (5.9) after expanding

$$T = \frac{1}{2\ell}\int_{-\sigma\ell}^{(1-\sigma)\ell} \rho(\sigma+\eta/\ell)\,(\dot{\mathbf{r}}^T\dot{\mathbf{r}} + 2\eta\,\dot{\mathbf{r}}^T\dot{\mathbf{b}} + \eta^2\,\dot{\mathbf{b}}^T\dot{\mathbf{b}})\,d\eta = \frac{1}{2}\,\dot{\mathbf{q}}^T(\mathbf{J}(\sigma)\otimes\mathbf{I})\,\dot{\mathbf{q}}.$$

Likewise, the angular momentum formula follows from

$$\mathbf{h} = \ell^{-1}\int_{\sigma\ell}^{(1-\sigma)\ell} \rho(\sigma+\eta/\ell)\,\eta^2\,(\tilde{\mathbf{b}}\,\dot{\mathbf{b}})\,d\eta = J\,\tilde{\mathbf{b}}\,\dot{\mathbf{b}}.$$

Generalized forces and torques

The matrix of generalized forces (5.21) is obtained after analyzing the work produced by the matrix of nodal forces $\mathbf{W}+\mathbf{F}$, where $\mathbf{W}$ is the matrix of external forces and $\mathbf{F}$ is the matrix of internal string forces (that is, the *ith* column of $\mathbf{W}$ summed with the *ith* column of $\mathbf{F}$ is the total force vector acting on node $\mathbf{n}_i$). Hence, from string connectivity, $\mathbf{F} = -\mathbf{N}\mathbf{C}_S^T\mathbf{\Gamma}\mathbf{C}_S$, and for an infinitesimal matrix displacement $\mathbf{\Delta_N}$

$$\text{trace}((\mathbf{W}+\mathbf{F})^T\mathbf{\Delta_N}) = \text{trace}(\mathbf{\Delta_N})^T(\mathbf{W}+\mathbf{F}) \;= \text{trace}\,\mathbf{\Delta_N}^T(\mathbf{W}-\mathbf{N}\mathbf{C}_S^T\mathbf{\Gamma}\mathbf{C}_S).$$

Recalling that $\mathbf{N} = \mathbf{Q}\,\mathbf{\Psi}^T + \mathbf{Y}$ so that $\mathbf{\Delta_N} = \mathbf{\Delta_Q}\mathbf{\Psi}^T$, then

$$\operatorname{trace}\mathbf{\Delta}_{\mathbf{N}}^T(\mathbf{W}+\mathbf{F}) = \operatorname{trace}\mathbf{\Psi}\mathbf{\Delta}_{\mathbf{Q}}^T(\mathbf{W}-\mathbf{N}\mathbf{C}_S^T\mathbf{\Gamma}\mathbf{C}_S) \quad = \operatorname{trace}\mathbf{\Delta}_{\mathbf{Q}}^T\mathbf{F}_{\mathbf{Q}},$$

where $\mathbf{F_Q} = (\mathbf{W} - \mathbf{N}\mathbf{C}_S^T\mathbf{\Gamma}\mathbf{C}_S)\mathbf{\Psi} = (\mathbf{W} - (\mathbf{Q}\,\mathbf{\Psi}^T + \mathbf{Y})\mathbf{C}_S^T\mathbf{\Gamma}\mathbf{C}_S)\mathbf{\Psi}$. In the absence of external forces, expression (5.43) follows from vectorization:

$$\mathbf{f_q} = \left(\mathbf{\Psi}^T \otimes \mathbf{I}_3\right)\mathbf{w} - \left(\mathbf{\Psi}^T\mathbf{C}_S^T\,\mathbf{\Gamma}\,\mathbf{C}_S \otimes \mathbf{I}_3\right)\left((\mathbf{\Psi}\otimes\mathbf{I}_3)\,\mathbf{q} + \mathbf{y}\right),$$

and using $\mathbf{\Phi} = \mathbf{\Psi}\otimes\mathbf{I}_3$.

Equations of motion

Newtonian approach The two equations that describe the constrained system are

$$J\,\tilde{\mathbf{b}}\,\ddot{\mathbf{b}} = \tau, \qquad\qquad \phi = \mathbf{b}^T\mathbf{b} - 1 = 0. \tag{5.52}$$

Here we will completely ignore the original constraint $\phi = 0$ and its first derivative, $\dot{\phi} = 0$, but we will honor the second derivative of the constraint $\ddot{\phi} = 0$. Assembling these into a single equation yields

$$\begin{bmatrix}\tilde{\mathbf{b}}\\ \mathbf{b}^T\end{bmatrix}\ddot{\mathbf{b}} = \begin{bmatrix}J^{-1}\tau\\ -\|\dot{\mathbf{b}}\|^2\end{bmatrix}. \tag{5.53}$$

The task of solving (5.53) for $\ddot{\mathbf{b}}$ is simply a linear algebra problem. Uniqueness of the solution is guaranteed by linear independent columns of the matrix coefficient of $\ddot{\mathbf{b}}$. We prove this linear independence by noting that

$$\begin{bmatrix}\tilde{\mathbf{b}}\\ \mathbf{b}^T\end{bmatrix}^T\begin{bmatrix}\tilde{\mathbf{b}}\\ \mathbf{b}^T\end{bmatrix} = \|\mathbf{b}\|^2\mathbf{I}.$$

The above identity can be expanded to provide an expression for the square of a skew-symmetric matrix

$$\tilde{\mathbf{b}}^2 = \mathbf{b}\mathbf{b}^T - \mathbf{b}^T\mathbf{b}\mathbf{I} = -\mathbf{P}(\mathbf{b})\|\mathbf{b}\|^2, \tag{5.54}$$

in terms of the projection matrix (5.27).

Next we use the properties of the unique Moore–Penrose inverse [SIG98] to conclude that the unique Moore–Penrose inverse of the coefficient matrix of the left-hand side of (5.53) is

$$\begin{bmatrix}\tilde{\mathbf{b}}\\ \mathbf{b}^T\end{bmatrix}^+ = \begin{bmatrix}-\tilde{\mathbf{b}} & \mathbf{b}\end{bmatrix}/\|\mathbf{b}\|^2.$$

Using this fact and $\tau = \tilde{\mathbf{b}}\sum_i^n \eta_i\mathbf{f}_i = \tilde{\mathbf{b}}\,\mathbf{f}_b$, the unique solution of (5.53) is

$$\begin{aligned}\ddot{\mathbf{b}} &= -J^{-1}\tilde{\mathbf{b}}\,\tau/\|\mathbf{b}\|^2 - \mathbf{b}(\|\dot{\mathbf{b}}\|/\|\mathbf{b}\|)^2\\ &= J^{-1}\mathbf{P}(\mathbf{b})\,\mathbf{f}_b - \mathbf{b}(\|\dot{\mathbf{b}}\|/\|\mathbf{b}\|)^2,\end{aligned}$$

where we have used (5.54).

Lagrangian approach From the Lagrangian function (5.29) and with the assumption that $f = 0$ the equations of motion of the rod are given by

$$m\,\ddot{\mathbf{r}} = \frac{d}{dt}\partial_{\dot{\mathbf{r}}}L = \partial_{\mathbf{r}}L = \mathbf{f}_{\mathbf{r}},$$

$$J\,\ddot{\mathbf{b}} = \frac{d}{dt}\partial_{\dot{\mathbf{b}}}L = \partial_{\mathbf{b}}L = \mathbf{f}_{\mathbf{b}} - J\xi\mathbf{b}, \tag{5.55}$$

$$\mathbf{0} = 2\,J^{-1}\partial_{\xi}L = \mathbf{b}^T\mathbf{b} - 1. \tag{5.56}$$

Now multiply (5.55) by $\mathbf{b}^T$ on the left and differentiate (5.56) twice with respect to time to obtain

$$J\,\mathbf{b}^T\ddot{\mathbf{b}} = \mathbf{b}^T\mathbf{f}_{\mathbf{b}} - J\xi\|\mathbf{b}\|^2, \qquad \mathbf{b}^T\ddot{\mathbf{b}} + \|\dot{\mathbf{b}}\|^2 = 0,$$

from where

$$\xi = (\|\dot{\mathbf{b}}\|/\|\mathbf{b}\|)^2 + J^{-1}\mathbf{b}^T\mathbf{f}_{\mathbf{b}}/\|\mathbf{b}\|^2.$$

Equation (5.32) follows after substituting ξ into (5.55) and using the definition of the projection matrix (5.27). Hence, the final form of the equations of motion is

$$m\,\ddot{\mathbf{r}} = \mathbf{f}_{\mathbf{r}},$$

$$J\,\ddot{\mathbf{b}} = \mathbf{P}(\mathbf{b})\mathbf{f}_{\mathbf{b}} - J(\|\dot{\mathbf{b}}\|/\|\mathbf{b}\|)^2\mathbf{b}. \tag{5.57}$$

Note that the two terms on the right-hand side of (5.57) are orthogonal to each other. This feature is expected to lend some efficiency to the numerical simulations, although we have not tried to quantify this.

5.6.2 Constrained Class 1 Tensegrity Structures

Note that when $\mathbf{\Phi} \neq \mathbf{\Psi} \otimes \mathbf{I}_3$ one should compute the vector of generalized forces $\mathbf{f}_{\mathbf{q}}$ using the vectorial version of the principle of virtual work

$$\delta_{\mathbf{n}}^T\,\mathbf{f} = \delta_{\mathbf{q}}^T\,\mathbf{t}_{\mathbf{q}},$$

after recalling that $\mathbf{n} = \mathbf{\Phi}\mathbf{q} + \mathbf{y}$ so that $\delta_{\mathbf{n}} = \mathbf{\Phi}\delta_{\mathbf{q}}$ and consequently

$$\delta_{\mathbf{n}}^T\,\mathbf{f} = \delta_{\mathbf{q}}^T\,\mathbf{\Phi}^T\mathbf{f} = \delta_{\mathbf{q}}^T\,\mathbf{t}_{\mathbf{q}},$$

where

$$\mathbf{t}_{\mathbf{q}} = \mathbf{\Phi}^T\mathbf{f} = -\mathbf{\Phi}^T(\mathbf{C}_S^T\mathbf{\Gamma}\mathbf{C}_S \otimes \mathbf{I}_3)(\mathbf{\Phi}\,\mathbf{q} + \mathbf{y}),$$

which is (5.43).

Single constrained rigid rod

Here is a proof that $\mathbf{M}(\mathbf{q})$ is nonsingular when $\mathbf{J} \succ 0$. Matrix $\mathbf{M}(\mathbf{q})$ is block lower-triangular; therefore, it is nonsingular whenever its diagonal blocks are nonsingular. In this case this means that $\mathbf{M}(\mathbf{q})$ is nonsingular if and only if its first diagonal block

$$\Sigma := m\,\mathbf{I} - f^2 J^{-1}\mathbf{E}^T\mathbf{P}(\mathbf{b})\mathbf{E}$$

is nonsingular. Recall that $\mathbf{q} \in \mathcal{Q}$ implies $\mathbf{b} \neq 0$ so that $\mathbf{P}(\mathbf{b})$ is well defined and that $\mathbf{E}$ is an orthonormal constant matrix, that is, it is full column rank and $\mathbf{E}^T\mathbf{E} = \mathbf{I}$. Therefore,

$$\Sigma = \mathbf{E}^T\Theta\mathbf{E}, \qquad \Theta := (m - f^2 J^{-1})\mathbf{I} + f^2 J^{-1}\mathbf{b}\mathbf{b}^T/\|\mathbf{b}\|^2.$$

Because $\mathbf{J} \succ 0$ we have that $J > 0$ and $m > f^2 J^{-1} \geq 0$. Therefore,

$$\Theta \succeq (m - f^2 J^{-1})\mathbf{I} \succ 0 \qquad \Longrightarrow \qquad \Sigma = \mathbf{E}^T\Theta\mathbf{E} \succ 0.$$

Hence $\mathbf{M}(\mathbf{q})$ is nonsingular.

Chapter 6

Closed-Loop Control of Tensegrity Structures

In this chapter we address the problem of designing closed loop control algorithms for tensegrity structures. In the literature, most closed-loop control algorithms for tensegrity structures have been developed for planar structures. This is understandable, since dynamic models for planar structures can be obtained using a minimal set of coordinates and ordinary differential equations (see [ASKD03, AS03]). In three dimensions, as shown in Chapter 5, one has to deal with differential-algebraic equations. No minimal ordinary differential equation model is possible. The options are to deal with singularities of the mass matrix, or to describe the system in a non-minimal set of coordinates without singularities, as done in Chapter 5. An additional nontrivial difficulty is to correctly model the strings, which are elements that cannot take compression. This can be thought of as a type of control saturation, which significantly complicates control design. In this chapter we present a control strategy for three-dimensional tensegrity structures that can address both of these issues.

The dynamic models are constructed using the same assumptions adopted in Chapter 5. In addition, we assume that the connectivity of the structure cannot be changed by the controller and that actuation is performed exclusively by changing the rest length of strings.

As for the control methodology, first a change-of-variables on the control inputs will be proposed that makes the nonlinear model affine on the control. Then a Lyapunov-based control design is introduced that can drive the system from an initial configuration toward the desired constant target coordinates satisfying the (string) control saturation constraints. The presentation closely follows that of [WdOS09].

R.E. Skelton, M.C. de Oliveira, *Tensegrity Systems*,
DOI 10.1007/978-0-387-74242-7_6, © Springer Science+Business Media, LLC 2009

6.1 Control of Tensegrity Systems

As in Chapter 5 we start by considering the simpler case of a single rigid rod.

6.1.1 A Single Rigid Rod

Let us first recall the form of the equations of motion obtained in Section 5.4 for general constrained class 1 tensegrity structures. We do this in the form of a lemma. The proof is in Section 5.4.

Lemma 6.1 *Let a rigid thin rod of mass* $m > 0$ *and length* $\ell > 0$ *be given as in Figure 5.1. Let the constant* $\sigma \in [0, 1]$ *be given so that the vector* $\mathbf{r} = \mathbf{v}(\sigma)$ *is a fixed point in the rod. Assume that* $\mathbf{r}$ *is subject to the linear constraint*

$$\mathbf{D}\,\mathbf{r} = \bar{\mathbf{r}} \tag{6.1}$$

where $r = \mathrm{rank}(\mathbf{D}) < 3$ *and* $\bar{\mathbf{r}} \in \mathbb{R}^r$ *constant. Let* $\mathbf{z} \in \mathbb{R}^{3-r}$ *and* $\mathbf{E} \in \mathbb{R}^{3\times(3-r)}$ *be an orthonormal matrix, i.e.,* $\mathbf{E}^T\mathbf{E} = \mathbf{I}$*, such that all solutions to the above equation are parametrized by*

$$\mathbf{r} = \mathbf{D}^{\dagger}\bar{\mathbf{r}} + \mathbf{E}\mathbf{z}.$$

Define the configuration vector

$$\mathbf{q} = \begin{pmatrix} \mathbf{z} \\ \mathbf{b} \end{pmatrix}, \tag{6.2}$$

and the orthogonal projection matrix

$$\mathbf{P}(\mathbf{b}) := \mathbf{I} - (\mathbf{b}\mathbf{b}^T)/\|\mathbf{b}\|^2. \tag{6.3}$$

The equations of motion governing the dynamics of the rigid rod are given by

$$\mathbf{M}(\mathbf{q})\,\ddot{\mathbf{q}} = \mathbf{g}(\mathbf{q}, \dot{\mathbf{q}}) + \mathbf{H}(\mathbf{q})\,\mathbf{t}_{\mathbf{q}}, \tag{6.4}$$

where

$$\mathbf{M}(\mathbf{q}) := \begin{bmatrix} m\,\mathbf{I} - f^2 J^{-1}\mathbf{E}^T\mathbf{P}(\mathbf{b})\mathbf{E} & \mathbf{0} \\ f J^{-1}\mathbf{P}(\mathbf{b})\mathbf{E} & \mathbf{I} \end{bmatrix},$$

$$\mathbf{H}(\mathbf{q}) := \begin{bmatrix} \mathbf{I} & -f J^{-1}\mathbf{E}^T\mathbf{P}(\mathbf{b}) \\ \mathbf{0} & J^{-1}\mathbf{P}(\mathbf{b}) \end{bmatrix},$$

$$\mathbf{g}(\mathbf{q}, \dot{\mathbf{q}}) := \begin{pmatrix} f(\|\dot{\mathbf{b}}\|/\|\mathbf{b}\|)^2\mathbf{E}^T\mathbf{b} \\ -(\|\dot{\mathbf{b}}\|/\|\mathbf{b}\|)^2\mathbf{b} \end{pmatrix}, \tag{6.5}$$

and $\mathbf{t}_{\mathbf{q}}$ *are the generalized forces acting on the rod written in the coordinate* $\mathbf{q}$.

Equation (6.4) is minimal with respect to the constraints against movement of the "pinned" coordinates, but they are not minimal with respect to the nonlinear constraints against changes in the bar length, as in (5.6). See Section 5.4 for details. The unconstrained (free rod) case, as in [SPM01, Ske05, dO06], can be obtained as a particular case of the above theorem where

$$\mathbf{z} = \mathbf{r}, \qquad \mathbf{E} = \mathbf{I}.$$

Indeed, in the unconstrained case, one can choose $\mathbf{r}$ to coincide with the center of mass of the rod without loss of generality, in which case $f = 0$ with much simplification to (6.4). In particular, the mass matrix $\mathbf{M}(\mathbf{q})$ becomes constant, as in [SPM01, Ske05, dO06].

In [WdOS09], we formally prove that one can ensure local existence and uniqueness of solutions to the above equations of motion when $\mathbf{t}_{\mathbf{q}}(t)$ is assumed to be a piecewise continuous function of t. It is also shown that constraint (5.6), which has been eliminated from the constrained equations of motion is indeed satisfied at the solutions of the unconstrained equations of motion (6.4) regardless of the forces $\mathbf{t}_{\mathbf{b}}$, which we will use to control the rod.

6.1.2 Control Inputs

A detailed discussion of string force is presented in Section 5.2.2. The Hookean assumption implies that the string stiffness is constant, so that control action is achieved by changing the rest length of the strings. For that matter we introduce the scalar control inputs

$$u_i := k_i(\|\mathbf{s_i}\| - l_i^0)\|/\|\mathbf{s_i}\| \geq 0, \qquad i = 1, \ldots, M. \tag{6.6}$$

The vector $\mathbf{s}_i$ denotes the ith string vector. Of course, for any $\mathbf{s}_i$ such that $\|\mathbf{s}_i\| \geq l_i^0$ there exists a corresponding $u_i \geq 0$. Conversely, for any $u_i \geq 0$ there exists a rest length l_i^0 and a material choice k_i to satisfy

$$l_i^0 = \|\mathbf{s_i}\|(1 - u_i/k_i).$$

However for a *fixed* material choice note that l_i^0 may be negative, which is not physically possible. However, in practice, the use of sufficiently stiff strings i.e., with $k_i \to \infty$, prevents this event from ever happening.

The main advantage of the above change-of-variable is that $\mathbf{t}_i$ as given in (5.18) can be replaced by the simpler expression

$$\mathbf{t}_i = -u_i\,\mathbf{s}_i, \qquad u_i \geq 0, \tag{6.7}$$

which is linear on the scalar control variable u_i that is subject to a positivity constraint. This feature will be important when deriving the control algorithm in the forthcoming sections. We assume that the strings can be made sufficiently stiff and long in order to realize the control laws to be obtained this way.

The change of variables (6.7) can be included in the generalized forces discussed in 5.2.1 by noticing that

$$\mathbf{t}_{\mathbf{q}} = -\Phi^T \sum_{i=1}^{M} (\mathbf{C}_{S_i}^T \otimes \mathbf{I}_3)\, \mathbf{s}_i\, u_i, = -\Phi^T \sum_{i=1}^{M} (\mathbf{C}_{S_i}^T \mathbf{C}_{S_i} \otimes \mathbf{I}_3)(\Phi\, \mathbf{q} + \mathbf{y})\, u_i. \quad (6.8)$$

Note that $\mathbf{t}_q$ is still affine on the transformed control vector $\mathbf{u}$. In the above formula, matrices $\boldsymbol{\Phi}$, $\mathbf{q}$ and $\mathbf{y}$ relates the node vector $\mathbf{n}$ to the configuration vector $\mathbf{q}$ through (5.42). Matrix $\mathbf{C}_S$ is the string connectivity matrix as defined in Sections 2.2 and 5.2.2.

6.1.3 General Class 1 Tensegrity Structures

Before proceeding with control design, we introduce notation that will simplify the exposition. Lemma 6.1 implies that the equations of motion of a single rod be written in the compact form

$$\mathbf{M}(\mathbf{q})\, \ddot{\mathbf{q}} = \mathbf{g}(\mathbf{q}, \dot{\mathbf{q}}) + \mathbf{H}(\mathbf{q})\, \mathbf{t}_{\mathbf{q}}$$

for some well-defined configuration vector $\mathbf{q}$. These equations of motion can be extended to cope with general class 1 tensegrity structures in a fairly straightforward way as done in Sections 5.4.2 and 5.4.2. There we have shown that the equations of motion for a general possibly constrained class 1 tensegrity system can be written as

$$\mathbf{M}_j(\mathbf{q}_j)\ddot{\mathbf{q}}_j = \mathbf{g}_j(\mathbf{q}_j, \dot{\mathbf{q}}_j) + \mathbf{H}_j(\mathbf{q})\, \mathbf{t}_{\mathbf{q}_j}, \qquad j = 1, \ldots, K,$$

where $\mathbf{M}_j$, $\mathbf{g}_j$, and $\mathbf{H}_j$ are as defined in Lemma 6.1 for the jth rod. Recall from Lemma 5.1 that $\mathbf{M}_j(\mathbf{q})$ is well defined and nonsingular when $\mathbf{J}_j \succ 0$. No changes are required on the string forces and, consequently, on the control inputs. Defining the augmented configuration vector

$$\mathbf{q} = \begin{pmatrix} \mathbf{q}_1^T & \cdots & \mathbf{q}_K^T \end{pmatrix}^T,$$

an associated constant matrix Φ and vector $\mathbf{y}$ can be constructed. Indeed, for a tensegrity system with M strings, one should still obtain an expression for the generalized force of the form (6.8), where $\mathbf{t}_{\mathbf{q}}$ now reflects the structure of the augmented configuration vector.

When developing control algorithms it is convenient to define the matrix

$$\mathbf{B}(\mathbf{q}) := \left[(\mathbf{C}_{S_1}^T \mathbf{C}_{S_1} \otimes \mathbf{I}_3)(\Phi\, \mathbf{q} + \mathbf{y}) \quad \cdots \quad (\mathbf{C}_{S_M}^T \mathbf{C}_{S_M} \otimes \mathbf{I}_3)(\Phi\, \mathbf{q} + \mathbf{y}) \right]. \quad (6.9)$$

The complete equations of motion can then be rewritten in the form

$$\mathbf{M}(\mathbf{q})\, \ddot{\mathbf{q}} = \mathbf{g}(\mathbf{q}, \dot{\mathbf{q}}) - \mathbf{H}(\mathbf{q})\, \Phi^T\, \mathbf{B}(\mathbf{q})\, \mathbf{u}, \qquad \mathbf{u} := (u_1, \cdots, u_M) \geq \mathbf{0}. \quad (6.10)$$

The above notation emphasizes the fact that dependence of the equations of motion on the transformed control input $\mathbf{u}$ is affine. The constraint on the control follows from the discussion in Section 6.1.2.

6.2 Lyapunov-Based Control Design

Our task is to design a controller for the non-minimal realization of the form (6.10). We focus on set-point control, that is, the task of designing control inputs that lead the system from some initial condition to a given constant set of coordinates. The results can be easily generalized to path-following control problems, which will not addressed in this book.

6.2.1 A Single Rigid Rod

Again to simplify the discussion consider the case of a single rigid rod. Define error vectors from the current coordinates $\mathbf{q} = (\mathbf{z}, \mathbf{b})$ to the desired target coordinates $\mathbf{q}_d = (\mathbf{z}_d, \mathbf{b}_d)$

$$\mathbf{e}_q := \begin{pmatrix} \mathbf{e}_z \\ \mathbf{e}_b \end{pmatrix}, \qquad \mathbf{e}_z := \mathbf{z} - \mathbf{z}_d, \qquad \mathbf{e}_b := \mathbf{b} - \mathbf{b}_d. \tag{6.11}$$

Consider a quadratic Lyapunov function candidate

$$V(\mathbf{q}, \dot{\mathbf{q}}) = V_z(\mathbf{z}, \dot{\mathbf{z}}) + V_b(\mathbf{b}, \dot{\mathbf{b}}) \tag{6.12}$$

where

$$\begin{aligned} V_z(\mathbf{z}, \dot{\mathbf{z}}) &= \frac{1}{2} \begin{pmatrix} \mathbf{e}_z \\ \dot{\mathbf{z}} \end{pmatrix}^T \begin{bmatrix} \mathbf{X}_z & \mathbf{Y}_z \\ \mathbf{Y}_z^T & \mathbf{Z}_z \end{bmatrix} \begin{pmatrix} \mathbf{e}_z \\ \dot{\mathbf{z}} \end{pmatrix}, \\ V_b(\mathbf{b}, \dot{\mathbf{b}}) &= \frac{1}{2} \begin{pmatrix} \mathbf{e}_b \\ \dot{\mathbf{b}} \end{pmatrix}^T \begin{bmatrix} \mathbf{X}_b & \mathbf{Y}_b \\ \mathbf{Y}_b^T & \mathbf{Z}_b \end{bmatrix} \begin{pmatrix} \mathbf{e}_b \\ \dot{\mathbf{b}} \end{pmatrix}. \end{aligned} \tag{6.13}$$

The term V_z involves only the translational coordinates and V_b only the rotational coordinates. If

$$\mathbf{Q}_z = \begin{bmatrix} \mathbf{X}_z & \mathbf{Y}_z \\ \mathbf{Y}_z^T & \mathbf{Z}_z \end{bmatrix} \succ 0, \qquad \mathbf{Q}_b = \begin{bmatrix} \mathbf{X}_b & \mathbf{Y}_b \\ \mathbf{Y}_b^T & \mathbf{Z}_b \end{bmatrix} \succ 0, \tag{6.14}$$

then $V_z(\mathbf{z}, \dot{\mathbf{z}}) > 0$ for all $\mathbf{z} \neq \mathbf{z}_d$ and $\dot{\mathbf{z}} \neq \mathbf{0}$ and $V_b(\mathbf{b}, \dot{\mathbf{b}}) > 0$ for all $\mathbf{b} \neq \mathbf{b}_d$ and $\dot{\mathbf{b}} \neq \mathbf{0}$. Therefore, $V(\mathbf{q}, \dot{\mathbf{q}}) > 0$ for all $\mathbf{q} \neq \mathbf{q}_d$ and $\dot{\mathbf{q}} \neq \mathbf{0}$.

As the coordinates $\mathbf{q}$ are not minimal, it is interesting to try to relate the Lyapunov function candidate (6.12) to some angular minimal coordinates. For instance, consider the three-dimensional entity $\mathbf{b}$ parametrized by two parameters θ and ϕ,

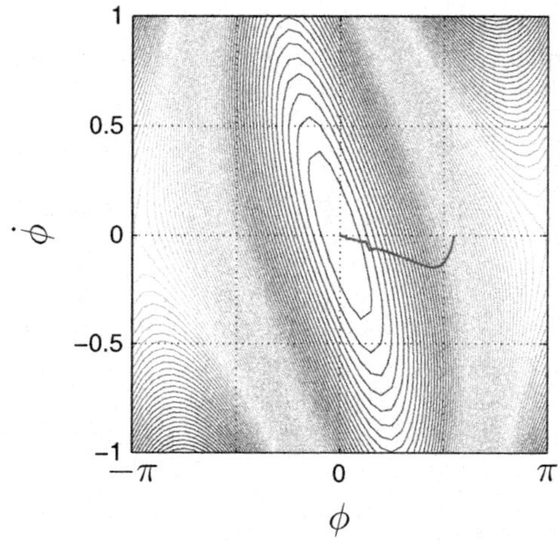

Figure 6.1: The Lyapunov landscape of $V_\phi(\phi, \dot{\phi})$ with system trajectory from Lyapunov-based control design with $\delta = 0.20$ and optimization of $\|\mathbf{u}\|_2$. This Lyapunov function uses $\alpha = \beta = 1$ and $\gamma = 3/2$

$$\mathbf{b} = \begin{pmatrix} \cos\theta \sin\phi \\ \sin\theta \sin\phi \\ \cos\phi \end{pmatrix}, \quad \|\mathbf{b}\| = 1,$$

$$\dot{\mathbf{b}} = \sin\phi\dot{\theta} \begin{pmatrix} -\sin\theta \\ \cos\theta \\ 0 \end{pmatrix} + \dot{\phi} \begin{pmatrix} \cos\theta\cos\phi \\ \sin\theta\cos\phi \\ -\sin\phi \end{pmatrix}, \quad \|\dot{\mathbf{b}}\| = \sqrt{\sin^2\phi\dot{\theta}^2 + \dot{\phi}^2},$$

where $\theta \in (-\pi, \pi]$ and $\phi \in [0, \pi]$. Now assume without loss of generality that $\mathbf{b}_d = (0, 0, 1)$ so that

$$\mathbf{b}_d^T \begin{bmatrix} \mathbf{b} & \dot{\mathbf{b}} \end{bmatrix} = \begin{bmatrix} \cos\phi & -\sin\phi\dot{\phi} \end{bmatrix}. \tag{6.15}$$

For illustration only, consider the particular choice $\mathbf{X}_b = \alpha\mathbf{I}$, $\mathbf{Y}_b = \beta\mathbf{I}$, $\mathbf{Z}_b = \gamma\mathbf{I}$, with $\alpha\gamma \geq \beta^2$, in which case

$$V_b(\mathbf{b}, \dot{\mathbf{b}}) = \frac{\alpha}{2}\|\mathbf{b} - \mathbf{b}_d\|^2 + \beta(\mathbf{b} - \mathbf{b}_d)^T\dot{\mathbf{b}} + \frac{\gamma}{2}\|\dot{\mathbf{b}}\|^2$$
$$= V_\phi(\phi, \dot{\phi}) + V_\theta(\phi, \dot{\theta}), \tag{6.16}$$

where

$$V_\phi(\phi, \dot{\phi}) = \alpha(1 - \cos\phi) + \beta\sin\phi\dot{\phi} + \frac{\gamma}{2}\dot{\phi}^2, \quad V_\theta(\phi, \dot{\theta}) = \frac{\gamma}{2}\sin^2\phi\dot{\theta}^2. \tag{6.17}$$

Notice the added complexity required to express the same Lyapunov function in a minimal coordinate system, here represented by $(\theta, \phi, \dot{\theta}, \dot{\phi})$. A contour plot of $V_\phi(\phi, \dot{\phi})$ is shown in Figure 6.1. Yet, the expressions (6.17) are possibly the simplest ones, a consequence of the particular choice $\mathbf{b}_d = (0, 0, 1)$. Any other choice in which the first and second components of $\mathbf{b}_d$ are not zero will make the resulting expressions significantly more involved. While from the point of view of simplifying V_b the choice $\mathbf{b}_d = (0, 0, 1)$ may seem a good idea that comes at the expense of extra complications. First, V_b is not a positive function of the state $(\theta, \phi, \dot{\theta}, \dot{\phi})$ (indeed, $V_b = 0$ for all $\phi = 0$ with θ and $\dot{\theta}$ arbitrary), which requires more involved stability analysis. Second, the mass matrix, which in minimal coordinates is a function of $(\theta, \phi, \dot{\theta}, \dot{\phi})$, is singular at $\phi = \phi_d = 0$, in which case a state space realization becomes numerically troubled near the intended equilibrium. All such concerns are absent in the control strategy to be presented in the sequel obtained with non-minimal coordinates.

6.2.2 A Control Design Problem

The next theorem shows how to compute $\mathbf{u}(t)$ as a function of time t based on the above Lyapunov function candidate in order to stabilize the tensegrity system.

Theorem 6.1 *Consider a system with a single rigid thin rod of mass $m > 0$, length $\ell > 0$, inertia $\mathbf{J} \succ 0$, and M controlled strings. Choose a configuration vector $\mathbf{q} = (\mathbf{z}, \mathbf{b})$ according to Lemma 6.1, depending on the constraints on the rod nodes. Let the matrix of nodes of the system be $\mathbf{N} \in \mathbb{R}^{3\times N}$, which locates the nodes where the rod and the strings connect to each other and to an inertial frame. Define the constant connectivity matrix $\mathbf{C}_S \in \mathbb{R}^{M\times N}$ which relates the matrix of string vectors $\mathbf{S} \in \mathbb{R}^{3\times M}$ to the node matrix through $\mathbf{S} = \mathbf{N}\mathbf{C}_S^T$, and the constant matrix Φ and constant vector $\mathbf{y}$ relating the configuration vector $\mathbf{q}$ to the matrix of nodes through* $\text{vector}\,\mathbf{N} = \Phi\,\mathbf{q} + \mathbf{y}$. *Define matrices $\mathbf{M}(\mathbf{q})$, $\mathbf{g}(\mathbf{q}, \dot{\mathbf{q}})$, and $\mathbf{H}(\mathbf{q})$ using the expressions (6.5) given in Lemma 6.1, and matrix $\mathbf{B}(\mathbf{q})$ as in (6.9). The motion of the system is governed by the differential equation*

$$\mathbf{M}(\mathbf{q})\,\ddot{\mathbf{q}} = \mathbf{g}(\mathbf{q}, \dot{\mathbf{q}}) - \mathbf{H}(\mathbf{q})\,\Phi^T\mathbf{B}(\mathbf{q})\,\mathbf{u}, \qquad \mathbf{u} \geq \mathbf{0}, \tag{6.18}$$

where $\mathbf{M}(\mathbf{q})$ is invertible for all $\mathbf{q}$. Let $\mathbf{q}_d = (\mathbf{z}_d, \mathbf{b}_d)$, $\mathbf{q}(0) \neq \mathbf{q}_d$, $\dot{\mathbf{q}}(0)$ and $\delta > 0$ be given, where $\|\mathbf{b}(0)\| = 1$. Assume that $\mathbf{q}_d$ is an equilibrium point of system (6.18), that is, there exists $\mathbf{u}_d \geq 0$ such that $\|\mathbf{b}_d\| = 1$ and $\mathbf{g}(\mathbf{q}_d, \mathbf{0}) = \mathbf{H}(\mathbf{q}_d)\,\Phi^T\mathbf{B}(\mathbf{q}_d)\,\mathbf{u}_d$. Select constant matrices $\mathbf{X}_b$, $\mathbf{Y}_b$, and $\mathbf{Z}_b$ so that $\mathbf{Z}_b \succ 0$ and $\mathbf{X}_b - \mathbf{Y}_b^Y\mathbf{Z}_b^{-1}\mathbf{Y}_b^T \succ 0$ ensuring $\mathbf{Q}_b \succ 0$. $\mathbf{X}_z$, $\mathbf{Y}_z$, and $\mathbf{Z}_z$ should satisfy the same criteria. Define $V(\mathbf{q}, \dot{\mathbf{q}})$ as in (6.12) and

$$\beta(\mathbf{q}, \dot{\mathbf{q}}) := \gamma(\mathbf{q}, \dot{\mathbf{q}}) + \mathbf{a}(\mathbf{q}, \dot{\mathbf{q}})^T\,\mathbf{g}(\mathbf{q}, \dot{\mathbf{q}}), \tag{6.19}$$

$$\mathbf{f}(\mathbf{q}, \dot{\mathbf{q}}) := \mathbf{B}(\mathbf{q})^T\Phi\,\mathbf{H}(\mathbf{q})^T\,\mathbf{a}(\mathbf{q}, \dot{\mathbf{q}}), \tag{6.20}$$

where

$$\gamma(\mathbf{q},\dot{\mathbf{q}}) := \begin{pmatrix} \mathbf{X}_z^T(\mathbf{z}-\mathbf{z}_d) + \mathbf{Y}_z^T\dot{\mathbf{z}} \\ \mathbf{X}_b^T(\mathbf{b}-\mathbf{b}_d) + \mathbf{Y}_b^T\dot{\mathbf{b}} \end{pmatrix}^T \dot{\mathbf{q}},$$

$$\mathbf{a}(\mathbf{q},\dot{\mathbf{q}}) := \mathbf{M}^{-T}(\mathbf{q}) \begin{pmatrix} \mathbf{Z}_z^T\dot{\mathbf{z}} + \mathbf{Y}_z^T(\mathbf{z}-\mathbf{z}_d) \\ \mathbf{Z}_b^T\dot{\mathbf{b}} + \mathbf{Y}_b^T(\mathbf{b}-\mathbf{b}_d) \end{pmatrix}. \quad (6.21)$$

If for all $t \geq 0$ there exists a control input $\mathbf{u}(t)$ such that

$$\mathbf{f}(\mathbf{q}(t),\dot{\mathbf{q}}(t))^T\mathbf{u}(t) = \beta(\mathbf{q}(t),\dot{\mathbf{q}}(t)) + \delta\, V(\mathbf{q}(t),\dot{\mathbf{q}}(t)), \qquad \mathbf{u}(t) \geq 0, \quad (6.22)$$

then $\lim_{t\to\infty}\mathbf{q}(t) = \mathbf{q}_d$.

A sketch of proof of the above theorem is given at the end of the chapter. See also [WdOS09]. Some remarks are in order. Several aspects of the above theorem present subtle technical difficulties. First, because of the presence of the inequality constraint in (6.22), it does not seem easy to guarantee that a feasible solution exists for all $t \geq 0$, therefore, leading the state asymptotically to the desired target. In order to characterize such property, one may need to impose restrictions on the initial conditions $(\mathbf{q}(0), \dot{\mathbf{q}}(0))$, the target configuration $\mathbf{q}_d$ and the rate of convergence δ. Nevertheless, as it will be illustrated by the examples in Section 6.3, the above construction is useful enough so that (6.22) remained feasible throughout the entire simulation time for the considered tensegrity structures. While this cannot be expected to be true in general, it should be pointed out that the initial conditions and target configurations were both feasible and stable tensegrity structures, that is, $\mathbf{q}(0)$ and $\mathbf{q}_d$ were selected such that

$$\exists\, \mathbf{u} \geq \mathbf{0}, \quad \mathbf{u} \neq \mathbf{0} \qquad \text{so that} \qquad \mathbf{g}(\mathbf{q},\dot{\mathbf{q}}) - \mathbf{H}(\mathbf{q})\,\Phi^T\mathbf{B}(\mathbf{q})\,\mathbf{u} = \mathbf{0},$$

and with an associated choice of material properties satisfying (6.6) made so as that $\mathbf{q}(0)$ and $\mathbf{q}_d$ were local minima of the total potential energy stored in the strings. These constraints are natural in tensegrity control, basically ensuring that one starts with a feasible tensegrity structure in static equilibrium that is taken to a different stable equilibrium configuration. Moreover, the results of the next section can at least be fully characterized when (6.22) has a solution or not.

On the theoretical side, the presence of the positivity constraint in (6.22) also significantly complicates the proof of Theorem 6.1. This is due to the fact that $\mathbf{u}(t)$ is not necessarily a continuous function, hence not Lipschitz, of the state $(\mathbf{q}, \dot{\mathbf{q}})$, as it would be the case if $\mathbf{u}(t)$ in (6.22) were unconstrained. Therefore, one cannot utilize a standard Lyapunov stability result such as [Kha96, Theorem 3.1]. Instead, one has to resort to more sophisticated tools that can handle differential equations with discontinuous right-hand sides, such as the ones described in [BR05]. Another contribution of the complete characterization of the solutions to (6.22) to be presented in the next section is the identification of a solution that is at least a continuous, although not differentiable, function of the state $(\mathbf{q}, \dot{\mathbf{q}})$. The proof at the end of the chapter covers just this simpler case. The results of the next section and some of the examples also consider solutions to (6.22) which are discontinuous.

One may wonder whether the choice of a more complex Lyapunov function V would be beneficial. While future research may prove this to be the case, as discussed previously, the function V, even though is a "simple" quadratic on the chosen set of non-minimal coordinates, gets fairly complicated if translated to a minimal set of coordinates. It is also reassuring that such a choice leads to a continuous control, when it is known that for the particular class of nonlinear systems that are affine on the control such control exists, as in [Art83].

Finally, as discussed in Section 6.1.3, the equations of motion for general class 1 tensegrity structures (6.10) is no different than equations of motion of a single rod (6.18). Indeed, if we define a combined Lyapunov function candidate of the form

$$V = \sum_{j=1}^{K} V_j$$

where each V_j is a Lyapunov function candidate written for each rod $j = 1, \dots, K$ as in (6.12), then Theorem 6.1 applies *verbatim* to entire tensegrity systems with more than one rod.

6.2.3 Admissible Control Inputs

Theorem 6.1 defines the constrained linear problem (6.22) that, if solvable for all $t \geq 0$, guarantees convergence to the target configuration $\mathbf{q}_d$. Some insight on the conditions for solvability can be obtained even before examining the complete solution. First note that if for some $t \geq 0$ we have $\beta + \delta V \neq 0$, then a necessary condition for solvability of (6.22) is that $\mathbf{a} \neq 0$. Of course, if $\beta + \delta V = 0$ then $\mathbf{u} = \mathbf{0}$ is a trivial solution, regardless of $\mathbf{a}$. The condition that $\mathbf{a} \neq \mathbf{0}$ requires that one selects $\mathbf{Y}_b \neq \mathbf{0}$ and $\mathbf{Y}_z \neq \mathbf{0}$ because

$$\mathbf{Y}_b = \mathbf{0}, \quad \mathbf{Y}_z = \mathbf{0}, \qquad \Longrightarrow \qquad \mathbf{a}(\mathbf{q}, \mathbf{0}) \equiv \mathbf{0} \quad \text{for all } \mathbf{q}.$$

This means that the system would be essentially "uncontrollable" from initial conditions with zero initial velocity. Even if this is not the case, i.e., $\dot{\mathbf{q}}(0) \neq \mathbf{0}$, it would likely create very high control inputs as the system approaches the target $\mathbf{q}_d$.

At a given time $t \geq 0$, the constrained control condition (6.22) may have multiple or no solutions. In the sequel we completely characterize solutions that minimize some norm of the control $\mathbf{u}(t)$. We consider three possibilities by minimizing the p-norms of the control input ($\|\mathbf{u}\|_p$) for the case $p = \{1, 2, \infty\}$. In all cases, explicit solutions for the constrained $\mathbf{u}(t)$ are available or can be proved not to exist. The problem we want to solve is, given $t \geq 0$ and $\mathbf{q}(t)$, $\dot{\mathbf{q}}(t)$, and $\delta > 0$, determine

$$\min_{\mathbf{u}(t)}\{\|\mathbf{u}(t)\|_p : \quad \mathbf{f}(\mathbf{q}(t), \dot{\mathbf{q}}(t))^T\mathbf{u}(t) = \beta\,(\mathbf{q}(t), \dot{\mathbf{q}}(t)) + \delta\, V(\mathbf{q}(t), \dot{\mathbf{q}}(t)), \quad \mathbf{u}(t) \geq 0\}, \tag{6.23}$$

where $\mathbf{f}$, β, and V are as defined in Theorem 6.1. The key is the following lemma, in which $\mathbf{v}_+$ denotes a vector where all negative entries of $\mathbf{v} \in \mathbb{R}^n$ are equal to zero. In other words, if $\mathbf{v}$ is an arbitrary vector in $\mathbb{R}^n$ then $\mathbf{v}_+$ is its two-norm projection on the positive orthant.

The next lemma is proved at the end of the chapter.

Lemma 6.2 *Let* $\mathbf{v} = (v_1, \ldots, v_n) \in \mathbf{R}^n$ *and* $\alpha \in \mathbf{R}$, $\alpha \neq 0$, *be given. The constrained optimization problem*

$$\mathbf{u}^p = \arg\min_{\mathbf{u}}\{\|\mathbf{u}\|_p : \quad \mathbf{v}^T\mathbf{u} = \alpha, \quad \mathbf{u} \geq \mathbf{0}\} \tag{6.24}$$

has a solution if and only if

$$\gamma := \max_i(\alpha^{-1}v_i) > 0. \tag{6.25}$$

If so, a solution is given by

(a) $\mathbf{u}^1 = (u_1, \cdots, u_n)$, *where* $u_i = 0$ *if* $i \neq i^*$ *and* $u_{i^*} = \gamma^{-1} > 0$ *for any* i^* *such that* $\alpha^{-1}v_{i^*} = \gamma$,

(b) $\mathbf{u}^2 = (\alpha^{-1}\mathbf{v})_+/\|(\alpha^{-1}\mathbf{v})_+\|_2^2$,

(c) $\mathbf{u}^\infty = (u_1, \cdots, u_n)$, *where* $u_i = 0$ *if* i *is such that* $\alpha^{-1}v_i \leq 0$ *and* $u_i = 1/[(\alpha^{-1}\mathbf{v})_+^T\mathbf{1}]$ *if* i *is such that* $\alpha^{-1}v_i > 0$,

for the values of $p = \{1, 2, \infty\}$.

The above lemma provides a solution to the control problem (6.23) by setting

$$\mathbf{v} = \mathbf{f}(\mathbf{q}(t), \dot{\mathbf{q}}(t)), \qquad \alpha = \beta(\mathbf{q}(t), \dot{\mathbf{q}}(t)) - \delta V(\mathbf{q}(t), \dot{\mathbf{q}}(t)).$$

Although the existence of a solution does not depend on p, different choices of p can yield vastly different control strategies. For example, the 1-norm solutions are realized using only one control (string) at a time, which may cause some undesired chattering between strings. On the other extreme, the ∞-norm solution applies equal force densities to all strings being actuated at time t. Both solutions yield in general, discontinuous controls $\mathbf{u}(t)$. In the $\mathbf{u}^1(t)$ case this happens because even though γ, $\mathbf{v}$ and α are continuous functions of $(\mathbf{q}, \dot{\mathbf{q}}(t))$, i^* will change abruptly depending on γ. In the $\mathbf{u}^\infty(t)$ case, a particular control channel u_i may suddenly switch from the positive value

$1/[(\alpha\mathbf{v})_+^T\mathbf{1}] > 0$ to 0 when $\alpha^{-1}v_i$ becomes non-positive. In contrast, $\mathbf{u}^2(t)$ is a continuous function of $(\mathbf{q}(t), \dot{\mathbf{q}}(t))$ because $\alpha^{-1}\mathbf{v}$ and its two-norm projection on the positive orthant are continuous functions of $(\mathbf{q}(t), \dot{\mathbf{q}}(t))$. Moreover, it is bounded whenever a solution exists since $\mathbf{v} = \mathbf{f}$ and α are bounded because $(\mathbf{q}(t), \dot{\mathbf{q}}(t))$ is bounded. Notice that the assumption that α is not zero can be imposed without loss of generality, since a small perturbation on δ can yield $\alpha \neq 0$ without compromising the control objectives.

The availability of closed formulas for the optimal control $\mathbf{u}^p$ at any time t means that this control can be readily implemented at very little computational cost provided that the state is available. The authors are currently investigating the design of state estimators which could be combined with the present results to construct output feedback controllers for tensegrity structures.

The control algorithm collapses if at some $t \geq 0$ it is found that $\gamma \leq 0$. In this case the forces needed to reduce the Lyapunov function cannot be realized by strings. In certain cases it may be possible to change the rate of convergence $\delta > 0$ in order to make $\gamma > 0$. In the numerical examples presented later in this chapter this problem never occurs. That shall be the case on most well-posed tensegrity problems where the choice of the desired target configuration, or a target trajectory, leads the tensegrity structure to a known stable equilibrium point.

Lemma 6.2 can be easily modified in order to find explicit expressions for the control inputs above certain threshold values. This may be important in cases where a residual level of pretension is required. Compare the above solution with the one for planar tensegrity structures proposed in [WdOS06a], which involves the solution of a much more complicated optimization problem.

6.3 Some Simple Examples

The theory developed so far is now illustrated by some simple examples.

Example 6.1

The simplest possible three-dimensional example of control of a class 1 tensegrity system is the case of a single rod pinned at one end actuated by three strings, as shown in Figure 6.2. A uniform mass distribution is assumed with

$$m = 1, \qquad \ell = 1, \qquad \sigma = 0 \qquad \Longrightarrow \qquad f = \frac{1}{2}, \qquad J = \frac{1}{3}.$$

The choice $\sigma = 0$ is made so that $\mathbf{r}$ coincides with one end of the rod which is fixed at the origin with a ball joint, i.e., $\mathbf{r} = \mathbf{0}$. The dynamics are described therefore by Lemma 6.1 with $\mathbf{q} = \mathbf{b}$. The string connectivity matrix $\mathbf{C}_S \in \mathbb{R}^{3\times 4}$

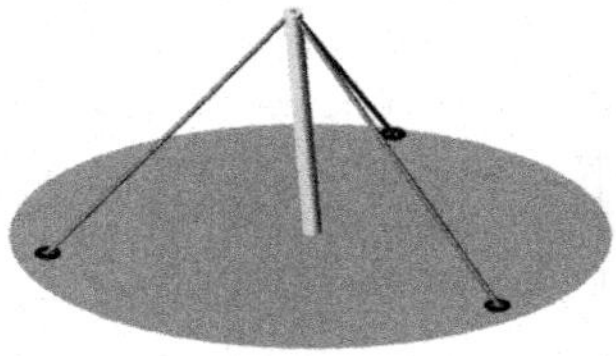

Figure 6.2: A single pinned rod and three strings

and the node matrix $\mathbf{N} \in \mathbb{R}^{3\times 4}$ are given by

$$\mathbf{C} = \begin{bmatrix}\mathbf{1} & -\mathbf{I}\end{bmatrix}, \qquad \mathbf{N} = \begin{bmatrix}\mathbf{n}_1 & \mathbf{n}_2 & \mathbf{n}_3 & \mathbf{n}_4\end{bmatrix}.$$

Matrix $\Phi \in \mathbb{R}^{12\times 3}$ and vector $\mathbf{y} \in \mathbb{R}^{12}$ as in (6.1.2) are

$$\Phi = \begin{bmatrix}\ell\mathbf{I} & \mathbf{0} & \mathbf{0} & \mathbf{0}\end{bmatrix}^T \qquad \mathbf{y} = \begin{pmatrix}\mathbf{0}^T & \mathbf{n}_2^T & \mathbf{n}_3^T & \mathbf{n}_4^T\end{pmatrix}^T.$$

The parameters of choice in the control design are the decay rate of the Lyapunov function, represented by the scalar δ, and the matrices $\mathbf{X}_b$, $\mathbf{Y}_b$, and $\mathbf{Z}_b$, all $\in \mathbb{R}^{3\times 3}$. Here

$$\delta = 0.15, \qquad \mathbf{X}_b = \mathbf{Y}_b = \mathbf{I}, \qquad \mathbf{Z}_b = \frac{3}{2}\mathbf{I}.$$

The initial and desired target configurations and the nodes $\mathbf{n}_i$ are

$$\mathbf{b}(0) = \frac{1}{\sqrt{3}}\begin{pmatrix}0\\ -\sqrt{2}\\ 1\end{pmatrix}, \quad \mathbf{b}_d = \frac{1}{\sqrt{5}}\begin{pmatrix}1\\ \sqrt{2}\\ \sqrt{2}\end{pmatrix},$$

$$\mathbf{n}_i = 1.2\begin{pmatrix}\cos[2\pi(i-2)/3]\\ \sin[2\pi(i-2)/3]\\ 0\end{pmatrix}, \qquad i = 2,\ldots,4$$

Nodes $\mathbf{n}_i$, $i = 2,\ldots,4$, represent fixed attachment points for the strings on the reference frame.

Figure 6.3 shows system trajectories from three simulations of the Lyapunov-based control design with different norm optimization, namely $\|\mathbf{u}\|_2$ (blue trajectory), $\|\mathbf{u}\|_1$ (green trajectory), and $\|\mathbf{u}\|_\infty$ (red trajectory). Detailed plots of the simulations including the trajectories, control input, and rod force (Lagrange multiplier), in the cases $p = 2$, $p = 1$, and $p = \infty$ are shown, respectively, in Figures 6.4, 6.5, and 6.6. The case $p = 1$ and $p = \infty$ exhibits discontinuity and chattering, particularly in the case $p = 1$. For illustration, the corresponding trajectory obtained in the case $p = 2$ is plotted on top of the contour plots of the Lyapunov function V_ϕ in Figure 6.1 in terms of the minimal coordinate angles $(\phi, \dot{\phi})$.

A more complex example is a class-1 tensegrity prism with 3 rods and 12 strings. Dynamics for such structure have been developed in Example 5.4. This example is developed next.

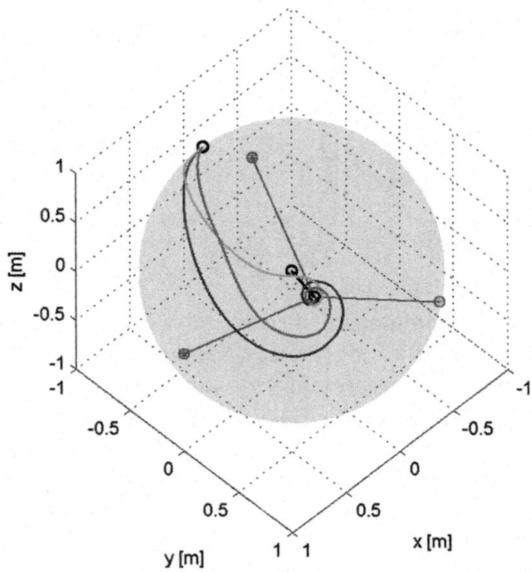

Figure 6.3: The system trajectories from simulations of the Lyapunov-based control design with $\delta = 0.15$ and optimization of $\|\mathbf{u}\|_2$, $\|\mathbf{u}\|_1$ and $\|\mathbf{u}\|_\infty$, colored *blue*, *green* and *red* respectively. The rod (*black*) is shown in the final configuration at t = 60s

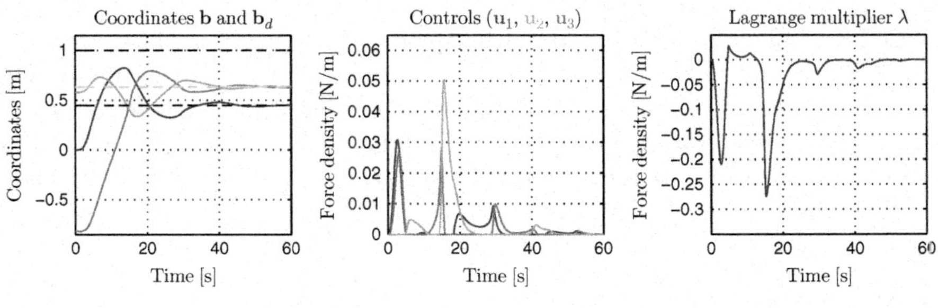

Figure 6.4: Data from simulation of the pinned single rod using the Lyapunov-based control design with $\delta = 0.15$ and optimization of $\|\mathbf{u}\|_2$. For coordinates *blue*, *green*, and *red* denote x-, y-, and z-components respectively. *Solid lines* denote actual values, while *dotted lines* denote desired constant target values. The *dashed black line* in the coordinates subplot denote $\|\mathbf{b}\|_2$. Further, the three strings are numbered from the one attached to position (1,0,0) and counterclockwise

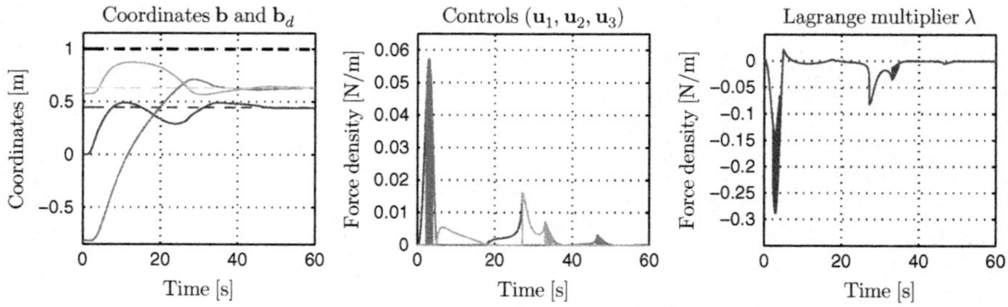

Figure 6.5: Data from simulation of the pinned single rod using the Lyapunov based control design with $\delta = 0.15$ and optimization of $\|\mathbf{u}\|_1$. For coordinates *blue*, *green*, and *red* denote x-, y-, and z-components, respectively. *Solid lines* denote actual values, while *dotted lines* denote desired constant target values. The *dashed black line* in the coordinates subplot denote $\|\mathbf{b}\|_2$. Further, the three strings are numbered from the one attached to position (1,0,0) and counterclockwise

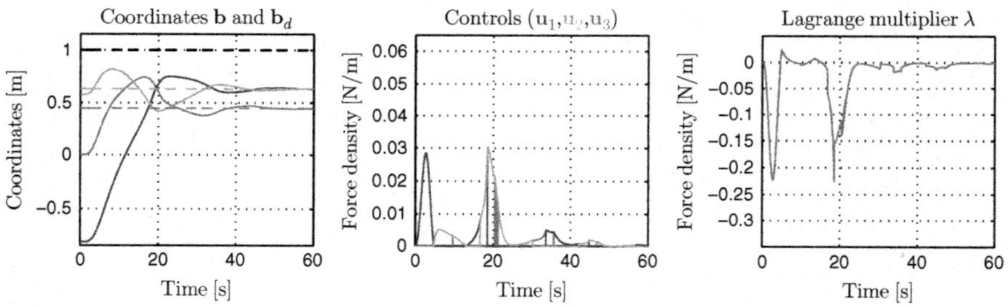

Figure 6.6: Data from simulation of the pinned single rod using the Lyapunov-based control design with $\delta = 0.15$ and optimization of $\|\mathbf{u}\|_\infty$. For coordinates *blue*, *green*, and *red* denote x-, y-, and z-components, respectively. *Solid lines* denote actual values, while *dotted lines* denote desired constant target values. The *dashed black line* in the coordinates subplot denote $\|\mathbf{b}\|_2$. Further, the three strings are numbered from the one attached to position (1,0,0) and counterclockwise

Example 6.2

Consider the tensegrity structure prism with three rods and twelve strings discussed in Examples 5.4 through 5.8 depicted in Figure 5.3. The control objective is to move from one stable configuration to another. The chosen parameters for the control design are the same for all three rods, that is,

$$\delta = 1, \qquad \mathbf{X}_{b_i} = \mathbf{Y}_{b_i} = \mathbf{X}_{r_i} = \mathbf{Y}_{r_i} = \mathbf{I},$$
$$\mathbf{Z}_{b_i} = \mathbf{Z}_{r_i} = \frac{3}{2}\mathbf{I}, \qquad i = \{1, 2, 3\}.$$

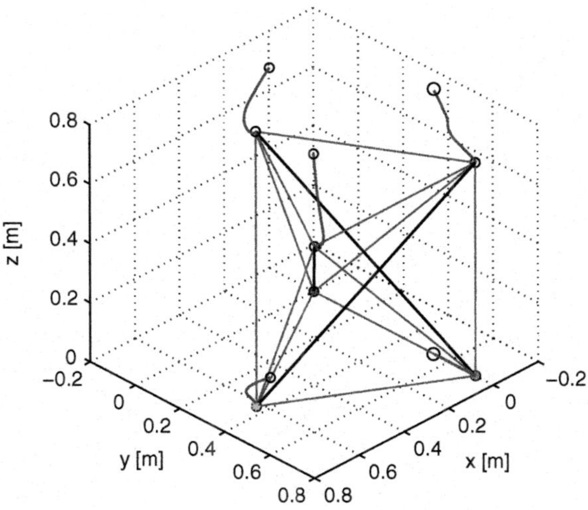

Figure 6.7: System trajectories from simulation of the tensegrity prism. The *blue lines* denote the trajectories of the nodes with the Lyapunov-based control design. The rods (*black*) are shown in their final configuration after 10s

The initial and desired target configurations for this structure are chosen so that both initial condition and target configuration are in static equilibrium. Such initial conditions are given by the initial vectors

$$\mathbf{z}_2(0) = \begin{pmatrix} 0.4659 \\ 0.2690 \end{pmatrix}, \qquad \mathbf{z}_3(0) = 0.5379,$$

$$\mathbf{b}_1(0) = \begin{pmatrix} 0.4243 \\ 0.4243 \\ 0.8000 \end{pmatrix}, \quad \mathbf{b}_2(0) = \begin{pmatrix} -0.5796 \\ 0.1553 \\ 0.8000 \end{pmatrix}, \quad \mathbf{b}_3(0) = \begin{pmatrix} 0.1553 \\ -0.5796 \\ 0.8000 \end{pmatrix},$$

associated with a symmetric and stable configuration of height equal to 0.8 m, and the desired target vectors

$$\mathbf{z}_{2d} = \begin{pmatrix} 0.6212 \\ 0.3586 \end{pmatrix} \qquad \mathbf{z}_{3d} = 0.7173,$$

$$\mathbf{b}_{1d} = \begin{pmatrix} 0.5657 \\ 0.5657 \\ 0.6000 \end{pmatrix}, \quad \mathbf{b}_2(0) = \begin{pmatrix} -0.7727 \\ 0.2071 \\ 0.6000 \end{pmatrix} \quad \mathbf{b}_{3d} = \begin{pmatrix} 0.2071 \\ -0.7727 \\ 0.6000 \end{pmatrix},$$

associated with a symmetric and stable configuration of height 0.6 m.

Figure 6.7 shows the trajectories of nodes for the Lyapunov-based control design (blue trajectories) for the tensegrity prism where the control norm is optimized for $p = 2$. Figure 6.8 presents essential data from the simulation of this system.

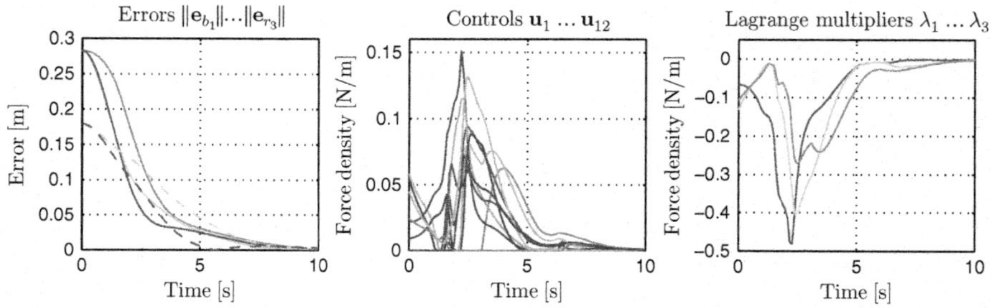

Figure 6.8: Data from simulation of the tensegrity prism with Lyapunov-based control design and optimization of $\|\mathbf{u}\|_2$. For coordinates *blue*, *green*, and *red* denote x-, y-, and z-components, respectively. For the error norms we use *blue*, *green*, and *red* for errors related to rod 1, 2, and 3, respectively. That gives $\|\mathbf{e}_{b_1}\|$, $\|\mathbf{e}_{b_2}\|$, $\|\mathbf{e}_{b_3}\|$, $\|\mathbf{e}_{r_1}\|$, $\|\mathbf{e}_{r_2}\|$, $\|\mathbf{e}_{r_3}\|$. Further, *Solid lines* denote the errors $\mathbf{b}_i$, i =1, 2 and 3, while *dashed lines* denote the errors $\mathbf{r}_i$, i =1, 2 and 3. The color code for control inputs, and strings, are u_1, u_2, u_3, u_4, u_5, u_6, and u_7. This combination is repeating for systems with more than seven inputs, such as this one

6.4 Chapter Summary

The control objective is to maneuver the system continuously from one configuration to another without any a priori information about a preferred trajectory. We have chosen a path that minimizes the total control magnitude at each step along the path. That is, the system continuously searches for the next step on a path toward the target configuration that minimizes some function of control inputs.

This control design is based on Lyapunov stability theory and yields in general a large degree of freedom to the optimization/allocation of control signals. For the tensegrity prims in Section 6.3, only one single scalar function needs to be matched by a combination of system inputs. The path from one configuration to another is therefore determined during the optimization problem. Algebraic constraints could be augmented to the optimization problem in order to decrease the inherent freedom of this design and to impact the choice of path toward the target configuration.

In Section 6.3, the convergence rate of the Lyapunov function was chosen to be exponentially stable, namely $\dot{V} = -\delta V$, $\delta > 0$. A sufficient requirement is $\dot{V} < 0$. The strict enforcement of this equality constraint also has a significant impact on the path chosen. An interesting feature, seen in Figures 6.4, 6.5, 6.6, and 6.8 is that even though the convergence of V is uniform, the convergence of system coordinates is not. The reason for this is the presence of algebraic constraints between system coordinates that need to be satisfied for all instants of time.

The choices of tunable parameter matrices $\mathbf{X}_b$, $\mathbf{Y}_b$, $\mathbf{Z}_b$, $\mathbf{X}_r$, $\mathbf{Y}_r$, and $\mathbf{Z}_r$ would naturally also alter the system trajectory. A general trend is that the system becomes slower and control inputs become larger with increasing diagonal elements of the matrices $\mathbf{Z}_z$ and $\mathbf{Z}_b$. The reason is that these elements penalize velocity, while V must decrease according to the convergence rate δ. Control inputs also become larger when reducing the diagonal terms of $\mathbf{Y}_z$ and $\mathbf{Y}_b$. This is because control inputs are directly multiplied by $\mathbf{Y}_z$ and $\mathbf{Y}_b$ in both designs, and that this term becomes dominant when $\dot{\mathbf{z}} \to \mathbf{0}$ and $\dot{\mathbf{b}} \to \mathbf{0}$. The system may become slower when increasing the diagonal terms of $\mathbf{Y}_z$ and $\mathbf{Y}_b$.

The optimization problem, which must be solved in order to provide a suitable combination of control inputs, has traditionally been solved using numerical iteration procedures. See, for instance [ASKD03, AS05, AS06, WdOS06a]. The proposed design presents explicit solutions to this problem, for all possible norms of the control inputs. An optimization of the two-norm presents a combination of continuous control signals where, in general, all strings gave a force contribution. An optimization of the one-norm gave a combination of control signals with discontinuities where only one string has a nonzero value at a given time. Also optimization of the infinity-norm gave a combination of control signals with discontinuities, here with all nonzero control signals having the same value.

There are several issues related to the implementation of these control laws, such as the constraint associated with the realizability of the control through strings being the most challenging one. In case the control problem proposed in this section has no admissible control solution, one possibility would be to use a passive control law that could be realized only with strings. One could switch into a passive mode while waiting for solvability of the control problem to be regained or a passive equilibrium is reached. Another source of concern is the loss of controllability at some configurations. For instance, in the pinned single rod example, if the rod vector is in the xy-plane, one has no control in the z-direction. How to avoid such singularities is still an open problem.

6.5 Advanced Material

6.5.1 Proof of Theorem 6.1

Use Lemma 6.1 and the discussion of the previous section to assemble the differential equation (6.18), which is affine on the control input $\mathbf{u} \geq \mathbf{0}$. First note that because $\mathbf{J} \succ 0$ we have $\mathbf{M}(\mathbf{q})$ is invertible (see Lemma 5.1). Therefore, system (6.18) can be written in state space form

$$\dot{\mathbf{x}} = \mathbf{k}(\mathbf{x}) - \mathbf{l}(\mathbf{x})\,\mathbf{u}, \qquad \mathbf{x} = \begin{pmatrix} \mathbf{q} \\ \dot{\mathbf{q}} \end{pmatrix},$$

$$\mathbf{k}(\mathbf{x}) = \begin{pmatrix} \dot{\mathbf{q}} \\ \mathbf{M}(\mathbf{q})^{-1}\mathbf{g}(\mathbf{q}) \end{pmatrix}, \qquad \mathbf{l}(\mathbf{x}) = \begin{pmatrix} \mathbf{0} \\ \mathbf{M}(\mathbf{q})^{-1}\mathbf{H}(\mathbf{q})\,\Phi^T\mathbf{B}(\mathbf{q}) \end{pmatrix},$$

where $\mathbf{k}$ and $\mathbf{l}$ are continuous functions of $\mathbf{x}$. In fact, functions $\mathbf{k}$ and $\mathbf{l}$ are continuously differentiable functions for all $\mathbf{x}$ such that $\mathbf{b} \neq \mathbf{0}$ (see [WdOS09, Lemma 1]). That implies $\mathbf{k}$ and $\mathbf{l}$ are locally Lipschitz for all $\mathbf{x}$ such that $\mathbf{b} \neq \mathbf{0}$. Clearly $V(\mathbf{q}, \dot{\mathbf{q}})$ is also continuously differentiable.

The main theoretical difficulty in the closed-loop stability analysis arises from the feedback control $\mathbf{u}(t)$ given in (6.22), which may not necessarily be a Lipschitz function of $\mathbf{q}$. However, as shown in Section 6.2.3, if (6.22) remains feasible up to sometime $T \geq 0$, then there exists a $\mathbf{u}(t) \geq \mathbf{0}$ satisfying (6.22) that is a continuous and bounded function of $\mathbf{q}(t)$ for $t \in [0, T]$. Continuity of $\mathbf{u}(t)$ can be used to ensure the local existence of solutions to the differential equations (6.18), while boundedness of $\mathbf{u}(t)$ can be used to construct a suitable version of Lyapunov stability, for instance by using the sample-and-hold ideas of [CLSS97]. Indeed, because $\mathbf{k}$ and $\mathbf{l}$ are Lipschitz, solutions $(\mathbf{q}(t), \dot{\mathbf{q}}(t))$ for a constant $\mathbf{u}(t) = \mathbf{u}(\tau)$ with $\mathbf{u}(\tau)$ bounded exist and are unique and continuous in $t \in [\tau, \tau + \delta]$ for some $\delta > 0$ (see [WdOS09, Lemma 4]).

Equipped with a proper Lyapunov stability theorem now verify that $V(\mathbf{q}_d, \mathbf{0}) = 0$. Furthermore, with (6.14), $V(\mathbf{q}, \dot{\mathbf{q}}) > 0$ for all $(\mathbf{q}, \dot{\mathbf{q}}) \neq (\mathbf{q}_d, \mathbf{0})$. Show that the time derivative of the Lyapunov function candidate (6.12) is

$$\dot{V}(\mathbf{q}, \dot{\mathbf{q}}, \ddot{\mathbf{q}}) = \gamma\,(\mathbf{q}, \dot{\mathbf{q}}) + \mathbf{a}(\mathbf{q}, \dot{\mathbf{q}})^T\mathbf{M}(\mathbf{q})\,\ddot{\mathbf{q}}.$$

Substitute $\mathbf{M}(\mathbf{q})\ddot{\mathbf{q}}$ from (6.18) in $\dot{V}$ above to obtain

$$\dot{V}(\mathbf{q}, \dot{\mathbf{q}}, \mathbf{u}) := \dot{V}(\mathbf{q},\, \dot{\mathbf{q}},\, \mathbf{g}(\mathbf{q}, \dot{\mathbf{q}}) - \mathbf{H}(\mathbf{q})\Phi^T\mathbf{B}(\mathbf{q})\mathbf{u}) = \beta\,(\mathbf{q}, \dot{\mathbf{q}}) - \mathbf{f}(\mathbf{q}, \dot{\mathbf{q}})^T\mathbf{u}.$$

Because (6.22) is assumed to be satisfied for all $t \geq 0$, use the above expression to show that

$$\dot{V}(\mathbf{q}(t), \dot{\mathbf{q}}(t), \mathbf{u}(t)) = -\delta\, V(\mathbf{q}(t), \dot{\mathbf{q}}(t)) < 0$$

for all $t \geq 0$. Moreover,

$$V(\mathbf{q}(t), \dot{\mathbf{q}}(t)) = e^{-\delta t}V(\mathbf{q}(0), \dot{\mathbf{q}}(0)) \implies \lim_{t\to\infty} V(\mathbf{q}(t), \dot{\mathbf{q}}(t)) = 0 \implies \lim_{t\to\infty} \mathbf{q} = \mathbf{q}_d.$$

if (6.22) is solvable for all $t \geq 0$.

6.5.2 Proof of Lemma 6.2

First let us prove the condition for the existence of a solution (6.25). For that we use Farkas' Lemma [BV04], which states that $\mathbf{A}\mathbf{x} = \mathbf{b}$ has no solution $\mathbf{x} \geq \mathbf{0}$ if and only if $\mathbf{A}^T\mathbf{y} \geq \mathbf{0}$ has some solution $\mathbf{b}^T\mathbf{y} < \mathbf{0}$. In the above problem, $\mathbf{A} = \alpha^{-1}\mathbf{v}^T$ and $\mathbf{b} = 1$ so that no solution exists if and only if

$$y\alpha^{-1}\mathbf{v} \geq \mathbf{0}, \qquad\qquad y < 0,$$

or, in other words, $\alpha^{-1}\mathbf{v} \leq \gamma\mathbf{1} \leq \mathbf{0}$, where $\mathbf{1} = (1, \cdots, 1)$. Therefore, γ must be positive for a solution to exist. The following auxiliary lemma will be used. A proof can be found in [WdOS09, Lemma 6].

Lemma 6.3 *Let $\mathbf{u}^*$ be the optimal solution to problem* (6.24). *Then $u_j^* = 0$ for all j such that $\alpha^{-1}v_j \leq 0$.*

Now consider the optimization problem in the 1-norm first, here reformulated as the linear program

$$\sigma = \min_{\mathbf{u}}\{\mathbf{1}^T\mathbf{u} : \quad \mathbf{u} \geq \mathbf{0}, \quad \alpha^{-1}\mathbf{v}^T\mathbf{u} = 1\}.$$

Assume that the above problem has some feasible solution, that is, $\gamma > 0$. In this case, its dual linear program

$$\rho = \max_{y}\{y : \quad y\alpha^{-1}\mathbf{v} \leq \mathbf{1}\}$$

should also have a solution. The dual problem, however, has a trivial solution $\rho = y^* = \gamma^{-1}$. From the dual solution a primal feasible optimal solution can be constructed as indicated in the lemma. Note that for $\mathbf{u} = \mathbf{u}^1 \geq \mathbf{0}$ we have $\sigma = \|\mathbf{u}^1\|_1 = \rho = \gamma^{-1}$.

We now reformulate problem (6.24) in the case $p = 2$ as

$$\mathbf{u}^2 = \arg\min_{\mathbf{u}}\{\frac{1}{2}\mathbf{u}^T\mathbf{u} : \quad (\alpha^{-1}\mathbf{v})_+^T\mathbf{u} = 1, \quad \mathbf{u} \geq \mathbf{0}\},$$

where $(\alpha^{-1}\mathbf{v})_+ \geq \mathbf{0}$. Because of Lemma 6.3, problem (6.24) and the above must have the same optimal solution. The above problem, in its turn, has the same solution as the relaxed optimization problem

$$\mathbf{u}^\# = \arg\min_{\mathbf{u}}\{\frac{1}{2}\mathbf{u}^T\mathbf{u} : \quad (\alpha^{-1}\mathbf{v})_+^T\mathbf{u} = 1\}$$

where $\mathbf{u}$ is not required to be nonnegative. Indeed, the optimality conditions for the above problem are such that

$$\mathbf{u}^\# = \lambda(\alpha^{-1}\mathbf{v})_+ \quad \Longrightarrow \quad (\alpha^{-1}\mathbf{v})_+^T\mathbf{u}^\# = \lambda\|(\alpha^{-1}\mathbf{v})_+\|_2^2 = 1$$
$$\Longrightarrow \quad \lambda = 1/\|(\alpha^{-1}\mathbf{v})_+\|_2^2.$$

Consequently

$$\mathbf{u}^2 = \mathbf{u}^{\#} = (\alpha^{-1}\mathbf{v})_+ / \|(\alpha^{-1}\mathbf{v})_+\|_2^2 \geq \mathbf{0}.$$

The case $p = \infty$ can be formulated as the linear program

$$(\zeta, \mathbf{u}^\infty) = \arg\min_{y,\mathbf{u}} \{ y : \quad \mathbf{0} \leq \mathbf{u} \leq y\mathbf{1}, \quad \alpha^{-1}\mathbf{v}^T\mathbf{u} = 1 \}.$$

Invoking Lemma 6.3 the above problem can be reformulated as

$$(\zeta, \mathbf{u}^\infty) = \arg\min_{y,\mathbf{u}} \{ y : \quad \mathbf{0} \leq \mathbf{u} \leq y\mathbf{1}, \quad (\alpha^{-1}\mathbf{v})_+^T\mathbf{u} = 1 \}.$$

For any feasible $\mathbf{u}$ and y we must have $(y\mathbf{1} - \mathbf{u}) \geq 0$ so that

$$0 \leq (\alpha^{-1}\mathbf{v})_+^T (y\mathbf{1} - \mathbf{u}) = y\,(\alpha^{-1}\mathbf{v})_+^T\mathbf{1} - 1 \quad \Longrightarrow \quad y \geq \frac{1}{(\alpha^{-1}\mathbf{v})_+^T\mathbf{1}}.$$

This bound is tight and equality holds, for instance, with $\mathbf{u}^\infty$ given in the theorem. By verifying that

$$(\zeta, \mathbf{u}^\infty) = (1/[(\alpha^{-1}\mathbf{v})_+^T\mathbf{1}], \mathbf{u}^\infty)$$

is a feasible solution to the original problem, one concludes optimality.

Bibliography

[Ald04] J. B. Aldrich. *Control Synthesis for a Class of Light and Agile Robotic, Tensegrity Structures.* PhD thesis, Department of Mechanical and Aerospace Engineering, University of California, San Diego, 2004.

[Art83] Z. Artstein. Stabilization with relaxed controls. *Nonlinear Analysis, Theory, Methods & Applications*, 7(11):1163–1173, 1983.

[AS03] J. B. Aldrich and R. E. Skelton. Control/stucture optimization approach for minimum-time reconfiguration of tensegrity systems. In *Proceedings of SPIE Smart Structures and Materials: Modeling, Signal Processing and Control*, volume 5049, pp. 448–459, San Diego, CA, USA, March 2003.

[AS05] J. B. Aldrich and R. E. Skelton. Time-energy optimal control of hyper-actuated mechanical systems with geometric path constraints. In *Proceedings of the 44th Conference on Decision and Control and the European Control Conference*, pp. 8246–8253, Seville, Spain, December 2005.

[AS06] J. B. Aldrich and R. E. Skelton. Backlash-free motion control of robotic manipulators driven by tensegrity motor networks. In *Proceedings of the 45th IEEE Conference on Decision and Control*, San Diego, CA, USA, December 2006.

[ASKD03] J. B. Aldrich, R. E. Skelton, and K. Kreutz-Delgado. Control synthesis for a class of light and agile robotic tensegrity structures. In *Proceedings of the American Control Conference*, pp. 5245–5251, Denver, CL, USA, June 2003.

[Ben89] M. P. Bendsoe. Optimal shape design as a material distribution problem. *Structural Optimization*, 1:193–202, 1989.

[Ben95] M. P. Bendsoe. *Optimization of Structural Topology, Shape, and Material.* Springer, 1995.

[BGFB94] S. P. Boyd, L. El Ghaoui, E. Feron, and V. Balakrishnan. *Linear Matrix Inequalities in System and Control Theory.* SIAM, Philadelphia, PA, USA, 1994.

[BK88] M. P. Bendsoe and N. Kikuchi. Generating optimal topologies in structural design using homogenization method. *Computer Methods in Applied Mechanics and Engineering*, 71:197–224, 1988.

[BR05] A. Bacciotti and L. Rosier. *Liapunov Functions and Stability in Control Theory.* Communications and Control Engineering. Springer, 2 edition, 2005.

[BS03] M. P. Bendsoe and O. Sigmund. *Topology Optimization: Theory, Methods and Applications.* Springer-Verlag, Berlin, 2003.

[BV04] S. P. Boyd and L. Vandenberghe. *Convex Optimization.* Cambridge University Press, 2004.

[CABS04] W. L. Chan, D. Arbelaez, F. Bossens, and R. E. Skelton. Active vibration control of a three-stage tensegrity structure. In K.W. Wang, editor, *Proceedings of the SPIE Smart Structures and Materials: Damping and Isolation*, volume 5386, pp. 340–346, San Diego, CA, USA, March 2004.

[Cal78] C. R. Calladine. Buckminster fuller's tensegrity structures and clerk maxwell's rules for the construction of stiff frames. *International Journal of Solids and Structures*, 14:161–172, 1978.

[Cav99] L. Caviglione. Modellazione statica e dinamica di una tensegrity nell'ambito di una strategia di controllo attivo. *Laurea Thesis, School of Engineering of the University of Genoa*, 1999.

[CB98] R. Connelly and A. Back. Mathematics and tensegrity. *American Scientist*, 86:142–151, 1998.

[CDGP] L. Caviglione, Andrea Del Grosso, and Mauro Pedretti. Stabilisation and control of large tensegrity structures. *Private communication.*

[Cha05] W. L. Chan. *Tensegrity Minimal Mass Design, Control and Experiments.* PhD thesis, Department of Mechanical and Aerospace Engineering, University of California, San Diego, 2005.

[CLO$^+$02] P. Cañadas, V. M. Laurent, C. Oddou, D. Isabey, and S. Wendling. A cellular tensegrity model to analyse the structural viscoelasticity of the cytoskeleton. *Journal of Theoretical Biology*, 218:155–173, 2002.

[CLSS97] F. H. Clarke, Y. S. Ledyaev, E. D. Sontag, and A. I. Subbotin. Asymptotic controllability implies feedback stabilization. *IEEE Transactions on Automatic Control*, 42(10):1394–1407, 1997.

[CP91] C. R. Calladine and S. Pellegrino. First-order infinitesimal mechanisms. *International Journal of Solids and Fluids*, 27(4):501–515, 1991.

[CS97] M. F. Coughlin and D. Stamenovic. A tensegrity structure with buckling compression elements: Application to cell mechanics. *Transactions of the ASME*, 64:480–486, 1997.

[CS02] W. Chan and R. E. Skelton. Equilibria and stiffness of planar tensegrity structures. In *AAS /AIAA Space Flight Mechanics Meeting*, San Antonio, Texas, January 2002.

[CW92] R. Connelly and W. Whiteley. The stability of tensegrity frameworks. *International Journal of Space Structures*, 7(2):153–163, 1992.

[CW96] R. Connelly and W. Whiteley. Second-order rigidity and prestress stability for tensegrity frameworks. *SIAM Journal of Discrete Mathematics*, 9(3):453–491, 1996.

[Dam92] A. A. Ten Dam. Stable numerical integration of dynamical systems subject to equality state-space constraints. *Journal of Engineering Mathemathics*, 26:315–337, 1992.

[DB92] A. R. Diaz and M. P. Bendsoe. Shape optimization of structures for multiple loading conditions using a homogenization method. *Structural Optimization*, 4:17–22, 1992.

[dJS01] B. de Jager and R. E. Skelton. Input/output selection for planar tensegrity models. In *Proceedings of the 40th IEEE Conference on Decision and Control*, pp. 4280–4285, Orlando, FL, USA, December 2001.

[dJSM01] B. de Jager, R. E. Skelton, and M. Masic. Integrated control/structure design for planar tensegrity models. In *Proceedings of the IEEE International Conference on Control Applications*, pp. 862–867, Glasgow, Scotland, UK, September 2001.

[DMPC98] S. Djouadi, R. Motro, J. C. Pons, and B. Crosnier. Active control of tensegrity systems. *ASCE Journal of Aerospace Engineering*, 11(2):37–44, 1998.

[dO05] M. C. de Oliveira. Dynamics of constrained tensegrity systems. Technical report, Internal Report, Dynamic Systems Research, Inc., San Diego, CA, USA, August 2005.

[dO06] M. C. de Oliveira. Dynamics of systems with rods. In *Proceedings of the 45th IEEE Conference on Decision and Control*, San Diego, CA, USA, December 2006.

[dOSC06] M. C. de Oliveira, R. E. Skelton, and W. Chan. Minimum mass design of tensegrity towers and plates. In *Proceedings of the 45th IEEE Conference on Decision and Control*, San Diego, CA, USA, December 2006.

[dOVV+09] M. C. de Oliveira, C. Vera, P. Valdez, Y. Sharma, R. E. Skelton, and L. A. Sung. Network nano-mechanics of the erythrocyte membrane skeleton in equibiaxial deformation. Submitted to the Biophysical Journal, 2009.

[dS03] T. d'Estrée Sterk. Using actuated tensegrity structures to produce a responsive architecture. In *Proceedings of the 2003 Annual Conference of the Association for Computer Aided Design In Architecture*, pp. 85–93, Indianapolis, IN, USA, October 2003.

[dS06a] T. d'Estrée Sterk. The office for robotic architectural media and the bureau for responsive architecture. www.oframbfra.com, 2006. Accessed September 4, 2006.

[dS06b] T. d'Estrée Sterk. Shape control in responsive architectural structures – current reasons and challenges. In *Proceedings of the 4th World Conference on Structural Control and Monitoring*, San Diego, CA, USA, July 2006.

[dS06c] T. d'Estrée Sterk. Smart buildings make smooth moves. www.wired.com. Accessed August 31, 2006.

[EG02] O. Egeland and J. T. Gravdahl. *Modeling and Simulation for Automatic Control*. Marine Cybernetics, Trondheim, Norway, 1 edition, 2002.

[Emm59] D. G. Emmerich. Carpentes perles (pearl framewroks). Institut National de la Propriété Industrielle (Registration 59423), May 1959.

[Emm96] D. G. Emmerich. Emmerich on self-tensioning structures. *International Journal of Space Structures*, 11:29–36, 1996.

[EP05] O. Ekeberg and K. Pearson. Computer simulation of stepping in the hind legs of the cat: An examination of mechanisms regulating the stance-to-swing transition. *Journal of Neurophysiology*, 94(6):4256–4268, 2005.

[EP06] O. Ekeberg and K. Pearson. Computer simulation of stepping in the hind legs of the cat: An examination of mechanisms regulating the stance-to-swing transition. *Journal of Neurophysiology*, 95(3):2028–2028, 2006.

[FMB+01] B. Fabry, G. M. Maksym, J. P. Butler, M. Glogauer, D. Navajas, and J. J. Fredberg. Scaling the microrheology of living cells. *Physical Review Letters*, 87:148102, 2001.

[Fos02] T. I. Fossen. *Marine Control Systems.* Marine Cybernetics, 2002.

[FS01] M. Fard and S. I. Sagatun. Exponential stabilization of a transversely vibrating beam via boundary control. *Journal of Sound and Vibration*, 240(4):613–622, 2001.

[Ful59] R. B. Fuller. US patent 3.063.521 Tensile integrity structures. United States Patent Office, 1959.

[Fur92] H. Furuya. Concept of deployable tensegrity structures in space applications. *International Journal of Space Structures*, 7(2):143–151, 1992.

[Gou98] M. Gough. In the laboratory of constructivism: Karl Ioganson's cold structures. *JSTOR*, 84:90–117, 1998.

[Han88] A. Hanaor. Prestressed pin-jointed structures – flexibility analysis and prestress design. *Computers and Structures*, 28:757–769, 1988.

[Han92a] A. Hanaor. Aspects of design of double-layer tensegrity domes *International Journal of Space Structures*, 7:101–113, 1992.

[Han92b] A. Hanaor. Double-layer tensegrity grids as deployable structures. *International Journal of Space Structures*, 8, 1992.

[Han94] A. Hanaor. Geometrically rigid double-layer tensegrity grids. *International Journal of Space Structures*, 9:227–238, 1994.

[HE08] N. Harischandra and O. Ekeberg. System identification of muscle-joint interactions of the cat hind limb during locomotion. *Biological Cybernetics*, 99(2):125–138, 2008.

[Hem73] W. S. Hemp. *Optimum Structures.* Oxford, Claredon Press, 1973.

[HJ85] R. Horn and C. A. Johnson. *Matrix Analysis.* Cambridge University Press, 1985.

[HP69] G. A. Hegemier and W. Prager. On Michell trusses. *International Journal of Mechanical Sciences*, 11(2):209–215, 1969.

[HP79] R. L. Huston and C. Passerello. On multi-body system dynamics. *Computers & Structures*, 10:439–446, 1979.

[Hug86] Peter C. Hughes. *Spacecraft Attitude Dynamics.* John Wiley & Sons, 1986.

[IJF06] I.-A. F. Ihle, J. Jouffroy, and T. I. Fossen. Formation control of marine surface craft: A Langrangian approach. *IEEE Journal of Oceanic Engineering*, 31:922–934, 2006.

[Ing98] D.E. Ingber. The architecture of life. *Scientific American*, pp. 48–57, January 1998.

[JKZ98] F. Jarre, M. Kocvara, and J. Zowe. Optimal truss design by interior point methods. *SIAM Journal on Optimization*, 8:1084–1107, 1998.

[JWL+07a] Ø. Jensen, A. S. Wroldsen, P. F. Lader, A. Fredheim, and M. A. Heide. Finite element analysis of tensegrity structures in offshore aquaculture installations. *Aquacultural Engineering*, 36(3):272–284, 2007.

[JWL+07b] Ø. Jensen, A. S. Wroldsen, P. F. Lader, A. Fredheim, M. A. Heide, and V. Johansen. Tensegrity structures in the design of flexible structures for offshore aquaculture. In *Proceedings of the 26th International Conference on Offshore Dynamics & Arctic Engineering (OMAE)*, San Diego, CA, USA, June 2007.

[Kha96] H. K. Khalil. *Nonlinear Systems (Third Edition).* Prentice-Hall Inc., 1996.

[KKAM99] K. Kebiche, M. N. Kazi-Aoual, and R. Motro. Geometrical non-linear analysis of tensegrity systems. *Journal of Engineering Structures*, 21(9):864–876, 1999.

[Kni00] B. F. Knight. *Deployable Antenna Kinematics using Tensegrity Structure Design.* Doctor of philosophy dissertation, University of Florida, Gainsville, 2000.

[Kob76] K. Kobyayashi. A biomedical study of human standing posture – a stability of standing posture against the external force. *Research Journal of Physical Education*, 20:257–268, 1976.

[KW02a] N. Kanchanasaratool and D. Williamson. Modeling and control of class nsp tensegrity structures. *International Journal of Control*, 75(2):123–139, 2002.

[KW02b] N. Kanchanasaratool and D. Williamson. Motion control of a tensegrity platform. *Communications in Information and Systems*, 2(3):299–324, 2002.

[Lak93] Roderic Lakes. Materials with structural hierarchy. *Nature*, 361:511–515, 1993.

[Lal96] H. Lalvani. Origins of tensegrity: views of Emmerich, Fuller and Snelson. *International Journal of Space Structures*, 11:27–55, 1996.

[Lan83] C. Lanczos. *The Variational Principles of Mechanics.* Dover Publications, Inc., 1983.

[Leo88] J. W. Leonard. *Tension Structures.* McGraw-Hill, 1988.

[Lie02] R. L. Lieber. *Skeletal Muscle Structure, Function and Plasticity.* Lippicott Williams & Wilkins, 2002.

[Lik69] P. W. Likins. Dynamics and control of flexible space vehicles. Technical Report JPL-TR-32-1329, Jet Propropulsion Lab, California Institute of Technology, Pasadena, CA, 1969.

[Luo04] H. Luo. *The Interaction of Near-Wall Turbulence with Compliant Tensegrity Fabrics: Modeling, Simulation, and Optimization.* PhD thesis, Department of Mechanical and Aerospace Engineering, University of California, San Diego, 2004.

[Mac65] Prof. Caroline MacGillavry. *Symmetry Aspects of M.C. Escher's Periodic Drawings*, 1965.

[Mal00] M. S. Malone. God, Stephen Wolfram, and everything else. *Forbes ASAP*, November 27, 2000.

[Mas04] M. Masic. *Design, Optimization, and Control of Tensegrity Structures.* Department of Mechanical and Aerospace Engineering, PhD thesis, University of California, San Diego, 2004.

[Max64] J. C. Maxwell. On the calculation of the equilibrium and stiffness of frames. *Phil. Mag.*, 27(294), 1864. Paper XXVI in Collected Papers, Cambridge, 1890.

[Maz03] E. Mazria. It's the architecture, stupid! *Solar Today*, pp. 48–51, May/June 2003.

[Mic04] A. G. M. Michell. The limits of economy in frame structures. *Philosophical Magazine*, 8:589–597, 1904.

[MM90] K. Miura and Y. Miyazak. Concept of the tension truss antenna. *AIAA Journal*, 28(6):1098–1104, 1990.

[MN01a] H. Murakami and Y. Nishimura. Initial shape finding and modal analyses of cycle right-cylindercal tensegrity modules. *Computers and Structures*, 79:891–917, 2001.

[MN01b] H. Murakami and Y. Nishimura. Static and dynamics characterization of some tensegrity modules. *Journal of Applied Mechanics*, 68:19–27, 2001.

[MNJ86] R. Motro, S. Najari, and P. Jouanna. Static and dynamic analysis of tensegrity systems. In *Proceedings of ASCE International Symposium on Shells and Spatial Structures, Computational Aspects*, pp. 270–279, New York, NY, USA, 1986, Springer.

[Mot03] R. Motro. *Tensegrity: Structural Systems for the Future.* Kogan Page Science, London, 2003.

[MR03] R. Motro and V. Raducanu. Tensegrity systems. *International Journal of Space Structures*, 18(2):77–84, 2003.

[MS02] M. Masic and R. E. Skelton. Deployable plates made from stable-element class 1 tensegrity. In A.M.R. McGowan, editor, *Proceedings of the SPIE Smart Structures and Materials: Industrial and Commercial Applications of Smart Structures Technologies*, volume 4698, pp. 220–230, San Diego, CA, USA, March 2002.

[MS04] M. Masic and R. E. Skelton. Open-loop control of class-2 tensegrity towers. In R.C. Smith, editor, *Proceedings of the SPIE Smart Structures and Materials: Modeling, Signal Processing, and Control*, volume 5383, pp. 298–308, San Diego, CA, USA, March 2004.

[MS05] Milenko Masic and Robert E. Skelton. Path planning and open-loop shape control of modular tensegrity structures. *Journal of Guidance, Control, and Dynamics*, 28(3): 421–430, 2005.

[MS06] M. Masic and R. E. Skelton. Selection of prestress for optimal dynamic/control performance of tensegrity structures. *International Journal of Solids And Structures*, 43(7–8):2110–2125, April 2006.

[MSG05] M. Masic, R. E. Skelton, and P. E. Gill. Algebraic tensegrity form-finding. *International Journal of Solids and Structures*, 42(16–17):4833–4858, August 2005.

[MSG06] M. Masic, R. E. Skelton, and P. E. Gill. Optimization of tensegrity structures. *International Journal of Solids and Structures*, 43(16):4687–4703, August 2006.

[MTBBS06] K. W. Moored, S. A. Taylor, T. K. Bliss, and H. Bart-Smith. Optimization of a tensegrity wing for biomimetic applications. In *Proceedings of the 45th IEEE Conference on Decision & Control*, San Diego, CA, USA, December 2006.

[Mur01] H. Murakami. Static and dynamic analyses of tensegrity structures, part 1. nonlinear equations of motion. *International Journal of Solids and Structures*, 38:3599–3613, 2001.

[OS01] M. Ostoja-Starzewski. Michell trusses in the presence of microscale material randomness: limitation of optimality. *Proceedings of the Royal Society of London Series A-Mathematical Physical and Engineering Sciences*, 457(2012):1787–1797, August 2001.

[OW97] I. J. Oppenheim and W. O. Williams. Tensegrity prisms as adaptive structures. In *ASME Annual Meeting*, Dallas, November 1997.

[OW00] I. J. Oppenheim and W. O. Williams. Geometric effects in an elastic tensegrity structure. *Journal of Elasticity*, 59:51–65, 2000.

[OW01a] I. J. Oppenheim and W. O. Williams. Vibration and damping in three-bar tensegrity structure. *Journal of Aerospace Engineering*, 14:85–91, 2001.

[OW01b] I. J. Oppenheim and W. O. Williams. Vibration of an elastic tensegrity structure. *European Journal of Mechanics and Solids*, 20:1023–1031, 2001.

[PC86] S. Pellegrino and C. R. Calladine. Matrix analysis of statically and kinematically indetermined frameworks. *International Journal of Solids and Structures*, 22:409–428, 1986.

[Pea90] P. Pearce. *Structure in Nature Is a Strategy for Design.* MIT Press, 1990, 5th printing.

[PEB06] K. Pearson, O. Ekeberg, and A. Buschges. Assessing sensory function in locomotor systems using neuro-mechanical simulations. *Trends in Neurosciences*, 29(11):625–631, 2006.

[Ped98] M. Pedretti. Smart tensegrity structures for the Swiss Expo. In *Smart Structures and Materials*, volume 3330, San Diego, CA, USA, 1998.

[Pel90] S. Pellegrino. Analysis of prestressed mechanisms. *International Journal of Solids and Structures*, 26:1329–1350, 1990.

[Pin05] J. P. Pinaud. *Deployable Tensegrity Towers.* PhD thesis, Department of Mechanical and Aerospace Engineering, University of California, San Diego, 2005.

[PMS03] J. P. Pinaud, M. Masic, and R. E. Skelton. Path planning for the deployment of tensegrity structures. In R. C. Smith, editor, *Proceedings of the SPIE Smart Structures and Materials: Modeling, Signal Processing, and Control*, volume 5049, pp. 436–447, San Diego, CA, USA, March 2003.

[PRLVC05] C. Paul, J. W. Roberts, H. Lipson, and F. J. Valero-Cuevas. Gait production in a tensegrity based robot. In *Proceedings of the International Conference on Advanced Robotics*, Seattle, WA, USA, July 2005.

[PSS04] J. P. Pinaud, S. Solari, and R. E. Skelton. Deployment of a class 2 tensegrity boom. In A. B. Flatau, editor, *Proceedings of the SPIE Smart Structures and Materials: Smart Structures and Integrated Systems*, volume 5390, pp. 155–162, San Diego, CA, USA, March 2004.

[Pug76] A. Pugh. *An Introduction to Tensegrity.* University of California Press, 1976.

[PVCL06] C. Paul, F.J. Valero-Cuevas, and H. Lipson. Design and control of tensegrity robots for locomotion. *IEEE Transactions on Robotics*, 22(5):944–957, October 2006.

[QKAM03] J. Quirrant, M. N. Kazi-Aoual, and R. Motro. Designing tensegrity systems: The case of a double layer grid. *Engineering Structures*, 25:1121–1130, 2003.

[Roz96] G. I. N. Rozvany. Some shortcomings in Michell's truss theory. *Structural Optimization*, 12(4):244–250, December 1996.

[Roz97] G. I. N. Rozvany. Some shortcomings in Michell's truss theory – author's reply and corrigendum. *Structural Optimization*, 13(2–3):203–204, April 1997.

[Roz98] G. I. N. Rozvany. Exact analytical solutions for some popular benchmark problems in topology optimisation. *Structural Optimization*, 15(1):42–48, February 1998.

[RW81] B. Roth and W. Whiteley. Tensegrity frameworks. *Transactions of the American Mathematical Society*, 265(2):419–446, 1981.

[Sad96] S. Sadao. Fuller on tensegrity. *International Journal of Space Structures*, 11:37–42, 1996.

[Sch81] L. A. Schmit. Structural synthesis – its genesis and development. *AIAA Journal*, 19(10):1249–1263, 1981.

[SCS98] C. Sultan, M. Corless, and R. E. Skelton. Nonlinear robust tracking control of a tensegrity motion simulator. In *Proceedings of the SPIE Conference on Acquisition, Tracking, and Pointing*, volume 3365, pp. 179–190, Orlando, FL, USA, April 1998.

[SCS99] C. Sultan, M. Corless, and R. E. Skelton. Peak to peak control of an adaptive tensegrity space telescope. In *Proceedings of the SPIE Conference on Mathematics and Control in Smart Structures*, volume 3667, pp. 190–201, Newport Beach, CA, USA, March 1999.

[SCS02] C. Sultan, M. Corless, and R. E. Skelton. Linear dynamics of tensegrity structures. *Engineering Structures*, 24:671–685, 2002.

[SHA+01a] R. E. Skelton, J. W. Helton, R. Adhikari, J. P. Pinaud, and W. L. Chan. An introduction to the mechanics of tensegrity structures. In *Proceedings of the 40th IEEE Conference on Decision and Control*, Orlando, FL, USA, December 2001.

[SHA+01b] R. E. Skelton, J. W. Helton, R. Adhikari, J. P. Pinaud, and W. L. Chan. *The Mechanical Systems Design Handbook: Modeling, Measurement, and Control*, An Introduction to the Mechanics of Tensegrity Structures. CRC Press, 2001.

[SIG98] R. E. Skelton, T. Iwasaki, and K. Grigoriadis. *A Unified Algebraic Approach to Linear Control Design.* Taylor & Francis, 1998.

[Ske05] R. E. Skelton. Dynamics and control of tensegrity systems. In *Proceedings of the IUTAM Symposium on Vibration Control of Nonlinear Mechanisms and Structures*, volume 130, Munich, Germany, July 2005.

[SMJ96] A. H. Simmons, C. A. Michal, and L. W. Jelinski. Molecular orientation and two-component nature of the crystalline fraction of dragline silk. *Science*, 271:84–87, 1996.

[Sne65] K. Snelson. US patent 3.169.611 Continuous tension discontinuous compression structures. United States Patent Office, 1965.

[Sne96] K. Snelson. Snelson on the tensegrity invention. *International Journal of Space Structures*, 11:43–48, 1996.

[SPM01] R. E. Skelton, J. P. Pinaud, and D. L. Mingori. Dynamics of the shell class of tensegrity structures. *Journal of the Franklin Institute*, 338(2–3):255–320, 2001.

[SPM02] R. E. Skelton, J. P. Pinaud, and D. L. Mingori. *The Mechanical Systems Design Handbook: Modeling, Measurement and Control*, The Dynamics of the Class 1 Shell Tensegrity Structure. CRC Press, 2002.

[SS97a] R. E. Skelton and C. Sultan. Controllable tensegrity, a new class of smart structures. In *Proceedings of the SPIE 4th Symposium on Smart Structures and Materials*, volume 3039, pp. 166–177, 1997.

[SS97b] C. Sultan and R. E. Skelton. Integrated design of controllable tensegrity structures. In *Proceedings of the ASME International Congress and Exposition*, volume 54, pp. 27–37, 1997.

[SS98a] C. Sultan and R. E. Skelton. Force and torque smart tensegrity sensor. In *Proceedings of SPIE Conference on Mathematics and Control in Smart Structures*, volume 3323, pp. 357–368, San Diego, CA, USA, March 1998.

[SS98b] C. Sultan and R. E. Skelton. Tendon control deployment of tensegrity structures. In *Proceedings of SPIE Conference on Mathematics and Control in Smart Structures*, volume 3323, pp. 455–466, San Diego, CA, USA, March 1998.

[SS03] R. E. Skelton and C. Sultan. Deployment of tensegrity structures. *International Journal of Solids and Structures*, 40:4637–4657, 2003.

[SS06] J. T. Scruggs and R. E. Skelton. Regenerative tensegrity structures for energy-harvesting applications. In *Proceedings of the 4th World Conference on Structural Control and Monitoring*, San Diego, CA, USA, July 2006.

[SSI04] C. Sultan, D. Stamenović, and D. E. Ingber. A computational tensegrity model predicts dynamic rheological behaviors in living cells. *Annals of Biomedical Engineering*, 32(4):520–530, 2004.

[Str61] Dirk Jan Struik. *Lectures on Classical Differential Geometry*. Addison-Wesley Pub. Co., Reading, Mass., 1961.

[Sul99] C. Sultan. *Modeling, design, and control of tensegrity structures with applications*. PhD thesis, Purdue University, School of Aeronautics and Astronautics, West Lafayette, IN, USA, 1999.

[Sul06] C. Sultan. Tensegrity structures research evolution. In *Proceedings of the 45th IEEE Conference on Decision and Control*, San Diego, CA, USA, December 2006.

[Tar80] T. Tarnai. Simultaneous static and kinematic interdeterminacy of space structures with cyclic symmetry. *International Journal of Solids and Structures*, 16:347–359, 1980.

[Ter94] Y. Termonia. Molecular modeling of spider silk elasticity. *Macromolecules*, 27:7378–7381, 1994.

[Tho00] M. W. Thomson. The AstroMesh deployable reflector. In S. Pellegrino and S. D. Guest, editors, *Proceedings of the IUTAM-IASS Symposium on Deployable Structures: Theory and Applications* (Cambridge, UK, September 1998), pp. 435–446, Dordrecht, The Netherlands, 2000, Kluwer Academic Publishers.

[Til02] G. Tilbert. *Deployable Tensegrity Structures for Space Applications.* Doctoral thesis, Royal Institute of Technology, Department of Mechanics, Stockholm, 2002.

[TP03] A. G. Tilbert and S. Pellegrino. Review of form-finding methods for tensegrity structures. *International Journal of Space Structures*, 18(4):209–223, 2003.

[Tri87] M. S. Triantafyllou. Dynamics of cables and chains. *The Shock and Vibration Digest*, 19(12):3–5, 1987.

[Tri91] M. S. Triantafyllou. Dynamics of cables, towing cables and mooring systems. *The Shock and Vibration Digest*, 23:3–8, 1991.

[Uit22] Bela Uitz. *Egység (1922). Reproduced in: The First Russian Show: A Commemoration of the Van Diemen Exhibition. Berlin 1922.* London: Annely Juda Fine Art, 1922.

[vdWdJ05] J. van de Wijdeven and B. de Jager. Shape change of tensegrity structures: Design and control. In *Proceedings of the American Control Conference*, pp. 2522–2527, Portland, OR, Portland, OR, USA, June 2005.

[VM99] N. Vassart and R. Motro. Multiparametered formfinding method: Application to tensegrity systems. *International Journal of Space Structures*, 14(2):147–154, 1999.

[VSBS05] C. Vera, R. E. Skelton, F. Bossens, and L. A. Sung. 3-D nanomechanics of an erythrocyte junctional complex in equibiaxial and anisotropic deformations. *Annals of Biomedical Engineering*, 33(10):1387–1404, 2005.

[WdOS06a] A. S. Wroldsen, M. C. de Oliveira, and R. E. Skelton. Configuration control of tensegrity structures. In *Proceedings of the 4th World Conference on Structural Control and Monitoring*, San Diego, CA, USA, July 2006.

[WdOS06b] A. S. Wroldsen, M. C. de Oliveira, and R. E. Skelton. A discussion on control of tensegrity systems. In *Proceeding of the 45th IEEE Conference on Decision and Control*, San Diego, December 2006.

[WdOS09] A. S. Wroldsen, M. C. de Oliveira, and R. E. Skelton. Modeling and control of non-minimal nonlinear realizations of tensegrity systems. *International Journal of Control*, 82(3): 389–407, 2009.

[Whi84] W. Whiteley. Infinitesimally rigid polyhedra. I. statics of frameworks. *Transactions of the American Mathematical Society*, 285(2):431–465, 1984.

[WJSS06] A. S. Wroldsen, V. Johansen, A. J. Sørensen, and R. E. Skelton. Hydrodynamic loading of tensegrity structures. In *Proceedings of the SPIE 13th Annual Symposium on Smart Structures and Materials: Modeling, Signal Processing, and Control*, San Diego, CA, USA, February 27–March 2 2006.

[WNS+01] N. Wang, K. Naruse, D. Stamenovic, J. J. Fredberg, S. M. Mijailovic, I. M. Tolić-Nørrelykke, T. Polte, R. Mannix, and D. E. Ingber. Mechanical behavior in living cells consistent with the tensegrity model. *Proceedings of the National Academy of Sciences*, 98(14):7765–7770, 2001.

[WOI99] S. Wendling, C. Oddou, and D. Isabey. Stiffening response of a cellular tensegrity. *Journal of Theoretical Biology*, 196:309–325, 1999.

[Wro07] A. S. Wroldsen. *Modelling and Control of Tensegrity Structures.* Ph.D. thesis, Norwegian University of Science and Technology, Department of Marine Technology, 2007.

[WRP+04] A. S. Wroldsen, A. M. Rustad, T. Perez, A. J. Sørensen, V. Johansen, P. F. Lader, A. Fredheim, and M. A. Heide. Tensegrity marine structure. Norwegian Patent NO 322560, US Patent US 2006102088, Chilean Patent CL2922-2005, Interational PCT Patent WO2005NO000425, 2004. Application Filed November 12, 2004.

[YP96] Z. You and S. Pelligrino. Cable-stiffened pantographic deployable structures part 1: Triangular mast. *AIAA Journal*, 34:813–820, 1996.

[ZMM06] L. Zhang, B. Maurin, and R. Motro. Form-finding of nonregular tensegrity systems. *Journal of Structural Engineering*, 132(9):1435–1440, 2006.

Index

Printed by Printforce, the Netherlands